Great Ideas of Science

A Reader in the Classic Literature of Science

Edited by
Robert M. Hazen
George Mason University, Fairfax, Virginia
Carnegie Institution of Washington, Geophysical Laboratory, Washington, DC

and
James Trefil
George Mason University, Fairfax, Virginia

San Diego, CA

Copyright © 2010 by Robert M. Hazen and James Trefil unless otherwise noted. All rights reserved. No part of this publication may be reprinted, reproduced, transmitted, or utilized in any form or by any electronic, mechanical, or other means, now known or hereafter invented, including photocopying, microfilming, and recording, or in any information retrieval system without the written permission of University Readers, Inc.

First published in the United States of America in 2010 by Cognella, a division of University Readers, Inc.

Trademark Notice: Product or corporate names may be trademarks or registered trademarks, and are used only for identification and explanation without intent to infringe.

14 13 12 11 10   1 2 3 4 5

Printed in the United States of America

ISBN: 978-1-935551-12-6

www.cognella.com 800.200.3908

CONTENTS

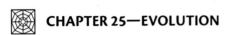 **CHAPTER 25—EVOLUTION**

PREFACE

SCIENTIFIC DISCOVERIES ARE remarkably varied in scope and content—made in the field, on the lab bench, or at the computer, with apparatus as sophisticated as a space-based telescope or as simple as a pencil and paper. But all of the discoveries of science are ultimately disseminated through the written word. For more than four centuries, in countless professional periodicals and technical treatises, the men and women of science have followed the same writing formula: What did I discover? How can you repeat what I did? What does it mean?

This volume was conceived as a companion to our textbook, *The Sciences: An Integrated Approach* (6th Edition, John Wiley & Sons, 2009). That text employs the "Great Ideas in Science" approach, which has been developed at George Mason University as an attempt to respond to the future needs of today's students. Our approach recognizes that science forms a seamless web of knowledge about the universe, and that a few overarching concepts (the "great ideas") unify all of the sciences—astronomy, biology, chemistry, geology and physics. Our goal is to serve the educational needs of people who will not be scientists but who need some knowledge of science to function as citizens.

Our central thesis in preparing this volume is that the core ideas of science are simple. Everyone lives in a world of matter and energy, forces and motions, and thus has daily experience in the workings of the Cosmos. We are all living beings and, likewise, have intimate familiarity with the principles of biology. Core ideas, therefore, provide a framework for our understanding of the universe—they give our scientific thinking structure and form. The great ideas represent a hierarchy in the sciences that transcend the boundaries of specific disciplines. Our organization around 25 central ideas, allows students to deal with the universe as it presents itself to them, rather than with the artificial disciplinary divisions that have arisen in academia. Our goal is to give each student the intellectual framework that will allow him or her to deal with the scientific aspects of problems that come into public debate.

Most of the following 50 excerpts were chosen to illustrate transformational discoveries in science history, such as the work of Copernicus, Newton, Darwin and Mendel. Other entries, including those of Snow, Cavendish, Van Helmont and Wöhler, expand on specific topics presented in *The Sciences*. Taken together, these readings reveal dramatic changes in the process and progress of science.

A central challenge of this volume has been to present portions of the original scientific texts in a way that is accessible to non-scientist readers. The language of science, especially in the past 100 years, often uses specialized vocabulary and high levels of mathematics to communicate to an audience of specialists. Indeed, most scientific papers today can be read and understood by only a tiny percentage of Ph.D. scientists, much less the non-scientific community. Nevertheless, most of the great scientific discoveries presented here are inherently simple and intuitive at their deepest level. Accordingly, we have edited most of the following papers to eliminate the most technical portions, and thus focus on the essence of the discoveries.

Throughout this volume you will share in these discoveries, as they were first presented to the public, and you will understand why we believe that science is the greatest ongoing adventure.

Robert M. Hazen & James Trefil
August, 2009

CHAPTER 1
SCIENCE AS A WAY OF KNOWING

Science is a way of asking and answering questions about the physical universe.

INTRODUCTION

THE SCIENTIFIC METHOD, by which observations lead to the development of a hypothesis, which makes predictions and can be tested by more observations, provides the foundation for discoveries about the natural world. Our first two excerpts, both taken from the medical literature, exemplify the scientific method in action.

William Harvey (1578–1677) was born in England and educated at Cambridge. He received his doctorate in medicine from the University of Padua in Italy, the premier medical institution in Europe at the time. He then returned to London, where he practiced medicine and lectured at the Royal College of Physicians. Harvey was one of the first people to apply the new technique of experiment and observation to medicine. In this book, he summarizes decades of experiments on the circulation of the blood. At the time, the primary authority on physiology was the Roman physician Galen, who taught that blood was created in the liver and sent from there to the cells, where it was absorbed. In the passages given below, Harvey overturned centuries of accepted dogma on this issue by pointing out that (1) the heart pumped too much blood for the circulation to be one way, and (2) observation showed that blood flowed out through the arteries and back through the veins. At that time, the connection between the arteries and veins wasn't known, although with the development of the microscope, anatomists discovered the network of capillaries that perform this function after his death.

On the Motion of the Heart and Blood in Animals[1]

by William Harvey (1628)

LETTER TO THE KING AND DEDICATION

To The Most Illustrious And Indomitable Prince Charles
King Of Great Britain, France, And Ireland Defender Of The Faith

Most Illustrious Prince!

THE HEART OF animals is the foundation of their life, the sovereign of everything within them, the sun of their microcosm, that upon which all growth depends, from which all power proceeds. The King, in like manner, is the foundation of his kingdom, the sun of the world around him, the heart of the republic, the fountain whence all power, all grace doth flow. What I have here written of the motions of the heart I am the more emboldened to present to your Majesty, according to the custom of the present age, because almost all things human are done after human examples, and many things in a King are after the pattern of the heart. The knowledge of his heart, therefore, will not be useless to a Prince, as embracing a kind of Divine example of his functions, and it has still been usual with men to compare small things with great. Here, at all events, best of Princes, placed as you are on the pinnacle of human affairs, you may at once contemplate the prime mover in the body of man, and the emblem of your own sovereign power. Accept therefore, with your wonted clemency, I most humbly beseech you, illustrious Prince, this, my new Treatise on the Heart; you, who are yourself the new light of this age, and indeed its very heart; a Prince abounding in virtue and in grace, and to whom we gladly refer all the blessings which England enjoys, all the pleasure we have in our lives.

Your Majesty's most devoted servant,
William Harvey.
London, 1628.

---

1   Translation by Alexander Bowie, London, 1889.

PREFATORY REMARKS

As we are about to discuss the motion, action, and use of the heart and arteries, it is imperative on us first to state what has been thought of these things by others in their writings, and what has been held by the vulgar and by tradition, in order that what is true may be confirmed, and what is false set right by dissection, multiplied experience, and accurate observation.

Almost all anatomists, physicians, and philosophers up to the present time have supposed, with Galen, that the object of the pulse was the same as that of respiration, and only differed in one particular, this being conceived to depend on the animal, the respiration on the vital faculty; the two, in all other respects, whether with reference to purpose or to motion, comporting themselves alike. …That it is blood and blood alone which is contained in the arteries is made manifest by the experiment of Galen, by arteriotomy, and by wounds; for from a single divided artery, as Galen himself affirms in more than one place, the whole of the blood may be withdrawn in the course of half an hour or less. The experiment of Galen alluded to is this: "If you include a portion of an artery between two ligatures, and slit it open lengthwise you will find nothing but blood"; and thus he proves that the arteries contain only blood.

1. Why, I ask, when we see that the structure of both ventricles is almost identical, there being the same apparatus of fibres, and braces, and valves, and vessels, and auricles, and both in the same way in our dissections are found to be filled up with blood similarly black in colour, and coagulated—why, I say, should their uses be imagined to be different, when the action, motion, and pulse of both are the same?

2. And, when we have these structures, in points of size, form, and situation, almost in every respect the same in the left as in the right ventricle, why should it be said that things are arranged in the former for the egress and regress of spirits, and in the latter or right ventricle, for the blood? The same arrangement cannot be held fitted to favour or impede the motion of the blood and of spirits indifferently. …

CHAPTER I: THE AUTHOR'S MOTIVES FOR WRITING

When I first gave my mind to vivisections, as a means of discovering the motions and uses of the heart, and sought to discover these from actual inspection, and not from the writings of others, I found the task so truly arduous, so full of difficulties, that I was almost tempted to think, with Fracastorius, that the motion of the heart was only to be comprehended by God.

At length, by using greater and daily diligence and investigation, making frequent inspection of many and various animals, and collating numerous observations, I thought that I had attained to the truth, that I should extricate myself and escape from this labyrinth, and that I had discovered what I so much desired, both the motion and the use of the heart and arteries. From that time I have not hesitated to expose my views upon these subjects, not only in private to my friends, but also in public, in my anatomical lectures, after the manner of the Academy of old.

CHAPTER IV: OF THE MOTION OF THE HEART AND ITS AURICLES
As Seen In The Bodies Of Living Animals

There are, as it were, two motions going on together: one of the auricles, another of the ventricles; these by no means taking place simultaneously, but the motion of the auricles preceding, that of the heart following; the motion appearing to begin from the auricles and to extend to the ventricles. When all things are becoming languid, and the heart is dying, as also in fishes and the colder blooded animals there is a short pause between these two motions, so that the heart aroused, as it were, appears to respond to the motion, now more quickly, now more tardily; and at length, when near to death, it ceases to respond by its proper motion, but seems, as it were, to nod the head, and is so slightly moved that it appears rather to give signs of motion to the pulsating auricles than actually to move.

But this especially is to be noted, that after the heart has ceased to beat, the auricles however still contracting, a finger placed upon the ventricles perceives the several pulsations of the auricles, precisely in the same way and for the same reason, as we have said, that the pulses of the ventricles are felt in the arteries, to wit, the distension produced by the jet of blood. And if at this time, the auricles alone pulsating, the point of the heart be cut off with a pair of scissors, you will perceive the blood flowing out upon each contraction of the auricles. Whence it is manifest that the blood enters the ventricles, not by any attraction or dilatation of the heart, but by being thrown into them by the pulses of the auricles.

CHAPTER V: OF THE MOTION, ACTION AND OFFICE OF THE HEART

From these and other observations of a similar nature, I am persuaded it will be found that the motion of the heart is as follows:

First of all, the auricle contracts, and in the course of its contraction forces the blood (which it contains in ample quantity as the head of the veins, the store-house and cistern of the blood) into the ventricle, which, being filled, the heart raises itself straightway, makes all its fibres tense, contracts the ventricles, and performs a beat, by which beat it immediately sends the blood supplied to it by the auricle into the arteries. The right ventricle sends its charge into the lungs by the vessel which is called vena arteriosa, but which in structure and function, and all other respects, is an artery. The left ventricle sends its charge into the aorta, and through this by the arteries to the body at large.

These two motions, one of the ventricles, the other of the auricles, take place consecutively, but in such a manner that there is a kind of harmony or rhythm preserved between them, the two concurring in such wise that but one motion is apparent, especially in the warmer blooded animals, in which the movements in question are rapid.

Even so does it come to pass with the motions and action of the heart, which constitute a kind of deglutition, a transfusion of the blood from the veins to the arteries. And if anyone, bearing these things in mind, will carefully watch the motions of the heart in the body of a living animal, he will perceive not only all the particulars I have mentioned, viz., the heart becoming erect, and making one continuous motion with its auricles; but farther, a certain obscure undulation and lateral inclination in the direction of the axis of the right ventricle, as if twisting itself slightly in performing its work.

The motion of the heart, then, is entirely of this description, and the one action of the heart is the transmission of the blood and its distribution, by means of the arteries, to the very extremities of the body; so that the pulse which we feel in the arteries is nothing more than the impulse of the blood derived from the heart.

CHAPTER VIII: OF THE QUANTITY OF BLOOD PASSING THROUGH THE HEART
From The Veins To The Arteries; And Of The Circular Motion Of The Blood

Thus far I have spoken of the passage of the blood from the veins into the arteries, and of the manner in which it is transmitted and distributed by the action of the heart; points to which some, moved either by the authority of Galen or Columbus, or the reasonings of others, will give in their adhesion. But what remains to be said upon the quantity and source of the blood which thus passes is of a character so novel and unheard-of that I not only fear injury to myself from the envy of a few, but I tremble lest I have mankind at large for my enemies, so much doth wont and custom become a second nature. Doctrine once sown strikes deep its root, and respect for antiquity influences all men. Still the die is cast, and my trust is in my love of truth and the candour of cultivated minds. And sooth to say, when I surveyed my mass of evidence, whether derived from vivisections, and my various reflections on them, or from the study of the ventricles of the heart and the vessels that enter into and issue from them, the symmetry and size of these conduits—for nature doing nothing in vain, would never have given them so large a relative size without a purpose—or from observing the arrangement and intimate structure of the valves in particular, and of the other parts of the heart in general, with many things besides, I frequently and seriously bethought me, and long revolved in my mind, what might be the quantity of blood which was transmitted, in how short a time its passage might be effected, and the like. But not finding it possible that this could be supplied by the juices of the ingested aliment without the veins on the one hand becoming drained, and the arteries on the other getting ruptured through the excessive charge of blood, unless the blood should somehow find its way from the arteries into the veins, and so return to the right side of the heart, I began to think whether there might not be a Motion, As It Were, In A Circle. Now, this I afterwards found to be true; and I finally saw that the blood, forced by the action of the left ventricle into the arteries, was distributed to the body at large, and its several parts, in the same manner as it is sent through the lungs, impelled by the right ventricle into the pulmonary artery, and that it then passed through the veins and along the vena cava, and so round to the left ventricle in the manner already indicated. This motion we may be allowed to call circular

And similarly does it come to pass in the body, through the motion of the blood, that the various parts are nourished, cherished, quickened by the warmer, more perfect, vaporous, spirituous, and, as I may say, alimentive blood; which, on the other hand, owing to its contact with these parts, becomes cooled, coagulated, and so to speak effete. It then returns to its sovereign, the heart, as if to its source, or to the inmost home of the body, there to recover its state of excellence or perfection. Here it renews its fluidity, natural heat, and becomes powerful, fervid, a kind of treasury of life, and impregnated with spirits, it might be said with balsam. Thence it is again dispersed. All this depends on the motion and action of the heart.

CHAPTER IX: THAT THERE IS A CIRCULATION OF THE BLOOD IS CONFIRMED

But lest anyone should say that we give them words only, and make mere specious assertions without any foundation, and desire to innovate without sufficient cause, three points present themselves for confirmation, which, being stated, I conceive that the truth I contend for will follow necessarily, and appear as a thing obvious to all. First, the blood is incessantly transmitted by the action of the heart from the vena cava to the arteries in such quantity that it cannot be supplied from the ingesta, and in such a manner that the whole must very quickly pass through the organ; second, the blood under the influence of the arterial pulse enters and is impelled in a continuous, equable, and incessant stream through every part and member of the body, in much larger quantity than were sufficient for nutrition, or than the whole mass of fluids could supply; third, the veins in like manner return this blood incessantly to the heart from parts and members of the body. These points proved, I conceive it will be manifest that the blood circulates, revolves, propelled and then returning, from the heart to the extremities, from the extremities to the heart, and thus that it performs a kind of circular motion.

Let us assume, either arbitrarily or from experiment, the quantity of blood which the left ventricle of the heart will contain when distended, to be, say, two ounces, three ounces, or one ounce and a half—in the dead body I have found it to hold upwards of two ounces. Let us assume further how much less the heart will hold in the contracted than in the dilated state; and how much blood it will project into the aorta upon each contraction; and all the world allows that with the systole something is always projected, a necessary consequence demonstrated in the third chapter, and obvious from the structure of the valves; and let us suppose as approaching the truth that the fourth, or fifth, or sixth, or even but the eighth part of its charge is thrown into the artery at each contraction; this would give either half an ounce, or three drachms, or one drachm of blood as propelled by the heart at each pulse into the aorta; which quantity, by reason of the valves at the root of the vessel, can by no means return into the ventricle. Now, in the course of half an hour, the heart will have made more than one thousand beats, in some as many as two, three, and even four thousand. Multiplying the number of drachms propelled by the number of pulses, we shall have either one thousand half ounces, or one thousand times three drachms, or a like proportional quantity of blood, according to the amount which we assume as propelled with each stroke of the heart, sent from this organ into the artery—a larger quantity in every case than is contained in the whole body!

But let it be said that this does not take place in half an hour, but in an hour, or even in a day; any way, it is still manifest that more blood passes through the heart in consequence of its action, than can either be supplied by the whole of the ingesta, or than can be contained in the veins at the same moment.

This truth, indeed, presents itself obviously before us when we consider what happens in the dissection of living animals; the great artery need not be divided, but a very small branch only (as Galen even proves in regard to man), to have the whole of the blood in the body, as well that of the veins as of the arteries, drained away in the course of no long time—some half-hour or less.

CHAPTER X: THE FIRST POSITION
Of The Quantity Of Blood Passing From The Veins To The Arteries, And That There Is A Circuit Of The Blood, Freed From Objections, And Farther Confirmed By Experiment

So far our first position is confirmed, whether the thing be referred to calculation or to experiment and dissection, viz., that the blood is incessantly poured into the arteries in larger quantities than it can be supplied by the food; so that the whole passing over in a short space of time, it is matter of necessity that the blood perform a circuit, that it return to whence it set out.

CHAPTER XI: THE SECOND POSITION IS DEMONSTRATED

That this may the more clearly appear to everyone, I have here to cite certain experiments, from which it seems obvious that the blood enters a limb by the arteries, and returns from it by the veins; that the arteries are the vessels carrying the blood from the heart, and the veins the returning channels of the blood to the heart; that in the limbs and extreme parts of the body the blood passes either immediately by anastomosis from the arteries into the veins, or mediately by the porosities of the flesh, or in both ways, as has already been said in speaking of the passage of the blood through the lungs whence it appears manifest that in the circuit the blood moves from that place to this place, and from the point to this one; from the centre to the extremities, to wit; and from the extreme parts back to the centre. Finally, upon grounds of calculation, with the same elements as before, it will be obvious that the quantity can neither be accounted for by the ingesta, nor yet be held necessary to nutrition.

Now let anyone make an experiment upon the arm of a man, either using such a fillet as is employed in blood-letting, or grasping the limb lightly with his hand, the best subject for it being one who is lean, and who has large veins, and the best time after exercise, when the body is warm, the pulse is full, and the blood carried in larger quantity to the extremities, for all then is more conspicuous; under such circumstances let a ligature be thrown about the extremity, and drawn as tightly as can be borne, it will first be perceived that beyond the ligature, neither in the wrist nor anywhere else, do the arteries pulsate, at the same time that immediately above the ligature the artery begins to rise higher at each diastole, to throb more violently, and to swell in its vicinity with a kind of tide, as if it strove to break through and overcome the obstacle to its current; the artery here, in short, appears as if it were preternaturally full. The hand under such circumstances retains its natural colour and appearance; in the course of time it begins to fall somewhat in temperature, indeed, but nothing is drawn into it.

After the bandage has been kept on for some short time in this way, let it be slackened a little, brought to that state or term of medium tightness which is used in bleeding, and it will be seen that the whole hand and arm will instantly become deeply coloured and distended, and the veins show themselves tumid and knotted; after ten or twelve pulses of the artery, the hand will be perceived excessively distended, injected, gorged with blood, drawn, as it is said, by this medium ligature, without pain, or heat, or any horror of a vacuum, or any other cause yet indicated.

If the finger be applied over the artery as it is pulsating by the edge of the fillet, at the moment of slackening it, the blood will be felt to glide through, as it were, underneath the finger; and he, too, upon whose arm the experiment is made, when the ligature is slackened, is distinctly conscious of a sensation of warmth, and of something, viz., a stream of blood suddenly making its way along the course of the vessels and diffusing itself through the hand, which at the same time begins to feel hot, and becomes distended.

From these facts it is easy for every careful observer to learn that the blood enters an extremity by the arteries; for when they are effectually compressed nothing is drawn to the member; the hand preserves its colour; nothing flows into it, neither is it distended; but when the pressure is diminished, as it is with the bleeding fillet, it is manifest that the blood is instantly thrown in with force, for then the hand begins to swell; which is as much as to say, that when the arteries pulsate the blood is flowing through them, as it is when the moderately tight ligature is applied; but where they do not pulsate, as, when a tight ligature is used, they cease from transmitting anything, they are only distended above the part where the ligature is applied.

On the Mode of Communication of Cholera

by John Snow, M.D.

Member Of The Royal College Of Physicians, Fellow Of The Royal Med. And Chir. Society, Fellow And Vice-President Of The Medical Society Of London

JOHN SNOW (1813–1858) was born in York, England, and attended medical school in London. He was, first and foremost, a practicing physician, and is best known in medical history as one of the pioneers in the use of anesthetics. He was, in fact, one of Queen Victoria's physicians, administering ether at several of her childbirths. From our point of view, however, his most important work involved his dogged gathering of data on the incidence of cholera during a series of outbreaks in London. Convinced that the disease was caused by something in the water its victims drank, he spent years tracing out London's hopelessly complex water supply systems, eventually coming to the point where he could associate each reported case of cholera with a specific contaminated water source. At that time, the only way water was tested was to see whether it looked clear and was odorless, a test met by even severely contaminated water. Although it wouldn't be until 1884 that scientists would identify the specific microorganism that caused cholera, Snow's patient collection of data eventually convinced the London authorities that they needed to pay attention to their water supply and led to our modern system of water purification.

Today, we regard Snow's relentless pursuit of data, even in the absence of a guiding theory, as one of the great historical examples of the scientific method in action.

* * *

ON THE MODE OF COMMUNICATION OF CHOLERA

The existence of Asiatic Cholera cannot be distinctly traced back further than the year 1769. Previous to that time the greater part of India was unknown to European medical men; and this is probably the reason why the history of cholera does not extend to a more remote period. It has been proved by various documents, quoted by Mr. Scot[1], that cholera was prevalent at Madras in the year above mentioned, and that it carried off many thousands of persons in the peninsula of India from that time to 1790. From this period we have very little account of the disease till 1814, although, of course, it might exist in many parts of Asia without coming under the notice of Europeans. ... There are certain circumstances, however, connected with the progress of cholera, which may be stated in a general way. It travels along the great tracks of human intercourse, never going faster than people travel, and generally much more slowly. In extending to a fresh island or continent, it always appears first at a seaport. It never attacks the crews of ships going from a country free from cholera, to one where the disease is prevailing, till they have entered a port, or had intercourse with the shore. Its exact progress from town to town cannot always be traced; but it has never appeared except where there has been ample opportunity for it to be conveyed by human intercourse. ...

Besides the facts above mentioned, which prove that cholera is communicated from person to person, there are others which show, first, that being present in the same room with a patient, and attending on him, do not necessarily expose a person to the morbid poison; and, secondly, that it is not always requisite that a person should be very near a cholera patient in order to take the disease, as the morbid matter producing it may be transmitted to a distance. ...

A consideration of the pathology of cholera is capable of indicating to us the manner in which the disease is communicated. If it were ushered in by fever, or any other general constitutional disorder, then we should be furnished with no clue to the way in which the morbid poison enters the system; whether, for instance, by the alimentary canal, by the lungs, or in some other manner, but should be left to determine this point by circumstances unconnected with the pathology of the disease. But from all that I have been able to learn of cholera, both from my own observations and the descriptions of others, I conclude that cholera invariably commences with the affection of the alimentary canal. ...

Diseases which are communicated from person to person are caused by some material which passes from the sick to the healthy, and which has the property of increasing and multiplying in the systems of the persons it attacks. ... As cholera commences with an affection of the alimentary canal, and as we have seen that the blood is not under the influence of any poison in the early stages of this disease, it follows that the morbid material producing cholera must be introduced into the alimentary canal-must, in fact, be swallowed accidentally, for persons would not take it intentionally; and the increase of the morbid material, or cholera poison, must take place in the interior of the stomach and bowels. ...

If the cholera had no other means of communication than those which we have been considering, it would be constrained to confine itself chiefly to the crowded dwellings of the poor, and would be continually liable to die out accidentally in a place, for want of the opportunity to reach fresh victims; but there is often a way open for it to extend itself more widely, and to reach the well-to-do classes of the community; I allude to the mixture of the cholera evacuations with the water used for drinking and culinary purposes, either by permeating the ground, and getting into wells, or by running along channels and sewers into the rivers from which entire towns are sometimes supplied with water. ...

The most terrible outbreak of cholera which ever occurred in this kingdom, is probably that which took place in Broad Street, Golden Square, and the adjoining streets, a few weeks ago. Within two hundred and fifty yards of the spot where Cambridge Street joins Broad Street, there were upwards of five hundred fatal attacks of cholera in ten days. The mortality in this limited area probably equals any that was ever caused in this country, even by the plague; and it was much more sudden, as the greater number of cases terminated in a few hours. The mortality would undoubtedly have been much greater had it not been for the flight of the population. Persons in furnished lodgings left first, then other lodgers went away, leaving their furniture to be sent for when they could meet with a place to put it in. Many houses were closed altogether, owing to the death of the proprietors; and, in a great number of instances, the tradesmen who remained had sent away their families: so that in less than six days from the commencement of the outbreak, the most afflicted streets were deserted by more than three-quarters of their inhabitants. There were a few cases of cholera in the neighbourhood of Broad Street, Golden Square, in the latter part of August; and the so-called outbreak, which commenced in the night between the 31st August and the 1st September was, as in all similar instances, only a violent increase of the malady. As soon as I became acquainted with the situation and extent of this irruption of cholera, I suspected some contamination of the water of the much-frequented street-pump in Broad Street, near the end of Cambridge Street; but on examining the water, on the evening, of the 3rd September, I found so little impurity in it of an organic nature, that I hesitated to come to a conclusion. Further inquiry, however, showed me that there was no other circumstance or agent common to the circumscribed locality in which this sudden increase of cholera occurred, and not extending beyond it, except the water of the above mentioned pump. I found, moreover, that the water varied, during the next two days, in the amount of organic impurity, visible to the naked eve, on close inspection, in the form of small white, flocculent particles; and I concluded that, at the commencement of the outbreak, it might possibly have been still more impure. I requested permission, therefore, to take a list, at the General Register Office, of the deaths from cholera, registered during the week ending 2nd September, in the subdistricts of Golden Square, Berwick Street, and St. Ann's, Soho, which was kindly granted. Eighty-nine deaths from cholera were registered, during the week, in the three subdistricts. Of these, only six occurred in the four first days of the week; four occurred on Thursday, the 31st August; and the remaining seventy-nine on Friday and Saturday. I considered, therefore, that the outbreak commenced on the Thursday; and I made inquiry in detail, respecting the eighty-three deaths registered as having taken place during the last three days of the week.

On proceeding to the spot, I found that nearly all the deaths had taken place within a short distance of the pump. There were only ten deaths in houses situated decidedly nearer to another street pump. In five of these cases the families of the deceased persons informed me that they always sent to the pump in Broad Street, as they preferred the water to that of the pump which was nearer. In three other cases, the deceased were children who went to school near the pump in Broad Street. Two of them were known to drink the water; and the parents of the third think it probable that it did so. The other two deaths, beyond the district which this pump supplies, represent only the amount of mortality from cholera that was occurring before the irruption took place.

With regard to the deaths occurring in the locality belonging to the pump, there were sixty-one instances in which I was informed that the deceased persons used to drink the pump-water from Broad Street, either constantly or occasionally. In six instances I could get no information, owing to the death or departure of every one connected with the deceased individuals; and in six cases I was informed that the deceased persons did not drink the pump-water before their illness.

The result of the inquiry then was, that there had been no particular outbreak or increase of cholera, in this part of London, except among the persons who were in the habit of drinking the water of the above-mentioned pump-web.

I had an interview with the Board of Guardians of St. James's parish, on the evening of Thursday, 7th September, and represented the above circumstances to them. In consequence of what I said, the handle of the pump was removed on the following day.

Besides the eighty-three deaths mentioned above as occurring on the three last days of the week ending September 2nd, and being registered during that week in the sub-districts in which the attacks occurred, a number of persons died in Middlesex and other hospital and a great number of deaths which took place in the locality during, the last two days of the week, were not registered till the week following. The deaths altogether, on the lst and 2nd of September, which have been ascertained to belong to this outbreak of cholera, were one hundred and ninety-seven; and many persons who were attacked about the same time as these, died afterwards. I should have been glad to inquire respecting the use of the water from Broad Street pump in all these instances, but was engaged at the time in an inquiry in the south districts of London, which will be alluded to afterwards and when I began to make fresh inquiries in the neighbourhood of Golden Square, after two or three weeks had elapsed, I found that there had been such a distribution of the remaining population that it would be impossible to arrive at a complete account of the circumstances. There is no reason to suppose, however, that a more extended inquiry would have yielded a different result from that which was obtained respecting the eighty-three deaths which happened to be registered within the district of the outbreak before the end of the week in which it occurred.

The additional facts that I have been able to ascertain are in accordance with those above related; and as regards the small number of those attacked, who were believed not to have drank the water from Broad Street pump, it must be obvious that there are various ways in which the deceased persons may have taken it without the knowledge of their friend. The water was used for mixing with spirits in all the public houses around. It was used likewise at dining-rooms and coffee-shops. The keeper of a coffee-shop in the neighbourhood, which was frequented by mechanics, and where the pump-water was supplied at dinner time, informed me (on 6th September) that she was already aware of nine of her customers who were dead. The pump-water was also sold in various little shops, with a teaspoonful of effervescing powder in it, under the name of sherbet; and it may have been distributed in various other ways with which I am unacquainted. The pump was frequented much more than is usual, even for a London pump in a populous neighbourhood.

There are certain circumstances bearing on the subject of this outbreak of cholera which require to be mentioned. The Workhouse in Poland Street is more than three-fourths surrounded by houses in which deaths from cholera occurred, yet out of five hundred and thirty-five inmates only five died of cholera, the other deaths which took place being those of persons admitted after in Broad Street, near to the pump, and on perceiving that no brewer's men were registered as having died of cholera, I called on Mr. Huggins, the proprietor. He informed me that there were above seventy workmen employed in the brewery, and that none of them had suffered from cholera, at least in a severe form, only two having been indisposed, and that not seriously at the time the disease prevailed. The men are allowed a certain quantity of malt liquor, and Mr. Huggins believes they do not drink water at all; and he is quite certain that the workmen never obtained water from the pump in the street. There is a deep well in the brewery, in addition to the New River water. …

I am indebted to Mr. Marshall for the following cases, which are interesting as showing the period of incubation, which in these three cases was from thirty-six to forty-eight hours. Mrs. — of 13 Bentinck Street, Berwick Street, aged 28, in the eighth month of pregnancy, went herself (although they were not usually water drinkers), on Sunday, 3rd September, to Broad Street pump for water. The family removed to Gravesend on the following day; and she was attacked with cholera on Tuesday morning at seven o'clock, and died of consecutive fever on 15th September, having been delivered. Two of her children drank also of the water, and were attacked on the same day as the mother, but recovered.

Dr. Fraser, of Oakley Square, kindly informed me of the following circumstance. A gentleman in delicate health was sent for from Brighton to see his brother at 6 Poland Street, who was attacked with cholera and died in twelve hours, on 1st September. The gentleman arrived after his brother's death, and did not see the body. He only stayed about twenty minutes in the house, where he took a hasty and scanty luncheon of rumpsteak, taking with it a small tumbler of brandy and water, the water being from Broad Street pump. He went to Pentonville, and was attacked with cholera on the evening of the following day, 2nd September, and died the next evening. …

In some of the instances, where the deaths are scattered a little further from the rest on the map, the malady was probably contracted at a nearer point to the pump. A cabinet-maker, who was removed from Philip's Court, Noel Street, to Middlesex Hospital, worked in Broad Street. A boy also who died in Noel Street, went to the National school at the end of Broad Street, and having to pass the pump, probably drank of the water. A tailor, who died at 6, Heddon Court, Regent Street, spent most of his time in Broad Street. A woman, removed to the hospital from 10, Heddon Court, had been nursing a person who died of cholera in Marshall Street. A little girl, who died in Ham Yard, and another who died in Angel Court, Great Windmill Street, went to the school in Dufour's Place, Broad Street, and were in the habit of drinking, the pump-water, as were also a child from Naylor's Yard, and several others who went to this and other schools near the pump in Broad Street. A woman who died at 2, Great Chapel Street, Oxford Street, had been occupied for two days preceding her illness at the public washhouses near the pump, and used to drink a good deal of water whilst at her work; the water drank there being sometimes from the pump and sometimes from the cistern. …

There is no doubt that the mortality was much diminished, as I said before, by the flight of the population, which commenced soon after the outbreak but the attacks had so far diminished before the use of the water was stopped, that it is impossible to decide whether the well still contained the cholera poison in an active state, or whether, from some cause, the water had become free from it. The pump-well has been opened, and I was informed by Mr. Farrell, the superintendent of the works, that there was no hole or crevice in the brickwork of the well, by which any impurity might enter; consequently in this respect the contamination of the water is not made out by the kind of physical evidence detailed in some of the instances previously related. I understand that the well is from twenty-eight to thirty feet in depth, and goes through the gravel to the surface of the clay beneath. The sewer, which passes within a few yards of the well is twenty-two feet below the surface. The water at the time of the cholera contained impurities of an organic nature, in the form of minute whitish flocculi visible on close inspection to the naked eve, as I before stated. Dr. Hassall, who was good enough to examine some of this water with the microscope, informed me that these particles had no organised structure, and that he thought they probably resulted from decomposition of other matter. He found a great number of very minute oval animalcules in the water, which are of no importance, except as an additional proof that the water contained organic matter on which they lived. The water also contained a large quantity of chlorides, indicating, no doubt, the impure sources from which the spring is supplied. Mr. Eley, the percussion-cap manufacturer of 37 Broad Street, informed me that he had long noticed that the water became offensive, both to the smell and taste, after it had been kept about two days. This, as I noticed before, is a character of water contaminated with sewage. Another person had noticed for months that a film formed on the surface of the water when it had been kept a few hours.

Whether the impurities of the water were derived from the sewers, the drains, or the cesspools, of which latter there are a number in the neighbourhood, I cannot tell. I have been informed by an eminent engineer, that whilst a cesspool in a clay soil requires to be emptied every six or eight months, one sunk in the gravel will often go for twenty years without being emptied, owing to the soluble matters passing away into the land-springs by percolation. As there had been deaths from cholera just before the great outbreak not far from this pump-well, and in a situation elevated a few feet above it, the evacuations from the patients might of course be amongst the impurities finding their way into the water, and judging the matter by the light derived from other facts and considerations previously detailed, we must conclude that such was the case. A very important point in respect to this pump-well is that the water passed with almost everybody as being perfectly pure, and it did in fact contain a less quantity of impurity than the water of some other pumps in the same parish, which had no share in the propagation of cholera. We must conclude from this outbreak that the

Date		No. of Fatal Attacks	Deaths
August	19	1	1
	20	1	0
	21	1	2
	22	0	0
	23	1	0
	24	1	2
	25	0	0
	26	1	0
	27	1	1
	28	1	0
	29	1	1
	30	8	2
	31	56	3
September	1	143	70
	2	116	127
	3	54	76
	4	46	71
	5	36	45
	6	20	37
	7	28	32
	8	12	30
	9	11	24
	10	5	18
	11	5	15
	12	1	6
	13	3	13
	14	0	6
	15	1	8
	16	4	6
	17	2	5
	18	3	2
	19	0	3
	20	0	0
	21	2	0
	22	1	2
	23	1	3
	24	1	0
	25	1	0
	26	1	2
	27	1	0
	28	0	2
	29	0	1
	30	0	0
Date Unknown		45	0
	Total	616	616

quantity of morbid matter which is sufficient to produce cholera is inconceivably small, and that the shallow pump-wells in a town cannot be looked on with too much suspicion, whatever their local reputation may be. Whilst the presumed contamination of the water of the Broad Street pump with the evacuations of cholera patients affords an exact explanation of the fearful outbreak of cholera in St. James's parish, there is no other circumstance which offers any explanation at all, whatever hypothesis of the nature and cause of the malady be adopted.

CHAPTER 2
THE ORDERED UNIVERSE

Newton's laws of motion and gravity predict the behavior of objects on Earth and in space.

INTRODUCTION

T H E CENTRAL TENET of science is that the universe is regular, predictable, and quantifiable. The following five excerpts illustrate more than 1500 years of progress in understanding the organization of the Cosmos.

Claudius Ptolemy lived in Alexandria about 100 AD (his exact dates are uncertain). At that time Alexandria was the center of intellectual life in the Mediterranean. A Greek speaking city ruling the country of Egypt, it has been compared by some scholars to a gated community, isolated by culture and language from its surroundings. Ptolemy was the director of the Museum, which you can think of as being something like a modern research laboratory or think tank. He collected data from hundreds of years of observations made by Greek and Babylonian astronomers and put them together into a complex model of the universe in which the earth was stationary at the center and the planets moved on spheres rolling within spheres. He put his work into a book that came to be called the *Almagest*. Translated into Arabic in the middle ages, the book came to Europe and was translated into Latin in the thirteenth century. It quickly became the primary astronomical text in Europe. Although Ptolemy's work was eventually supplanted by the Copernican system and later by the clockwork universe of Isaac Newton, the following short excerpt from the *Almagest* is typical of the kind of reasoning he used. It also shows clearly that educated people knew that the Earth was round well before the time of Columbus.

Almagest
by Claudius Ptolomy (c. 100 AD)

THAT ALSO THE EARTH, TAKEN AS A WHOLE, IS SENSIBLY SPHERICAL

Now THAT ALSO the earth taken as a whole is sensibly spherical, we could most likely think out in this way. For again it is possible to see that the sun and moon and the other stars do not rise and set at the same time for every observer on the earth, but always earlier for those living towards the orient and later for those living towards the occident. For we find that the phenomena of eclipses taking place at the same time, especially those of the moon, are not recorded at the same hours for everyone—that is, relatively to equal intervals of time from noon; but we always find later hours recorded for observers towards the orient than for those towards the occident. And since the differences in the hours is found to be proportional to the distance between the place, one would reasonably suppose the surface of the earth spherical, with the result that the general uniformity of curvature would assure every part's covering those following it proportionately. But this, would not happen if the figure were any other, as can be seen from the following considerations.

For, if it were concave, the rising stars would appear first to people towards the occident; and if it were flat, the stars would rise and set for all people together and at the same time; and if it were a pyramid, a cube, or any other polygonal figure, they would again appear at the same time for all observers on the same straight line. But none of these things appears to happen. It is further clear that it could not be cylindrical with the curved surface turned to the risings and set tings and the plane bases to the poles of the universe, which some think more plausible. For then never would any of the stars be always visible to any of the inhabitants of the curved surface, but either all the stars would both rise and set for observers or the same stars for an equal distance from either of the poles would always be invisible to all observers. Yet the more we advance towards the north pole, the more the southern stars are hidden and the northern stars, appear. So it is clear that here the curvature of the earth covering parts uniformly in oblique directions proves its spherical, form on every side. Again, whenever we sail towards mountains or any high places from whatever angle and in what ever direction, we see their bulk little by little increasing as if they were arising from the sea, whereas before they seemed submerged because of the curvature of the water's surface.

On the Revolutions of the Celestial Spheres
by Nicolas Copernicus
Nuremburg, 1543

NICOLAS COPERNICUS (1473–1542) was born in Torun. He studied at the Jagellonian University in Krakow before taking up canon law and medicine in Italy. He returned to Poland to take up a position as the business manager for a large cathedral. In that position, he was responsible for administering the extensive land and property holding of the Church. In addition, he was active in Polish politics, serving on a commission to reform the currency and leading several military expeditions. Think of him as the rough equivalent of the mayor of a large city or the governor of a small state today. Like most educated men of his time, he had an intellectual hobby—astronomy in his case. He did very little observing himself, but used the data in Ptolemy's *Almagest* to ask a fundamental question: was it possible to construct a model in which the Earth moved and the sun stood still? Responding to repeated requests from friends and colleagues, he finally did publish his model which, while not anything like our current picture, was the first serious attempt to present a heliocentric model of the universe. This preface, which many scholars believe was written by theologian Andreas Osiander rather than by Copernicus himself, shows how a savvy church politician was able to publish his work without running into the kinds of problems that later plagued Galileo.

* * *

TO THE READER CONCERNING THE HYPOTHESES OF THIS WORK

Since the newness of the hypotheses of this work—which sets the earth in motion and puts an immovable sun at the centre of the universe—has already received a great deal of publicity, I have no doubt that certain of the savants have taken grave offense and think it wrong to raise any disturbance among liberal disciplines which have had the right set-up for a long time now. If, however, they are willing to weigh the matter scrupulously, they will find that the author of this work has done nothing which merits blame. For it is the job of the astronomer to use painstaking and skilled observation in gathering together the history of the celestial movements, and then—since he cannot by any line of reasoning reach the true causes of these movements—to think-up or construct whatever causes or hypotheses he pleases such that, by the assumption these causes, those same movements can be calculated from the principles of geometry for the past and for the future too. This artist is markedly outstanding in both of these respects: for it is not necessary that these hypotheses should be true, or even, probable; but it is enough if they provide a calculus which fits the observations—unless by some chance there is anyone so ignorant of geometry and optics as to hold the epicycle of Venus as probable and to believe this to be a cause why Venus alternately proceeds and follows the sun at angular distance of up to 40° or more. For who does not see necessarily follows from this assumption that the diameter of the planet in as perigee should appear more than four times greater and the body of the planet more than sixteen times greater; than in its apogee? Nevertheless the experience of all the ages is opposed to that. There are also other things in this discipline which are just as absurd, but it is not necessary to examine them right now. For it is sufficiently clear that this art is ignorant of the cause of the apparent irregular movements.

But since for one and the same movement varying hypotheses are proposed from time to time, as eccentricity or epicycle for the movement of the sun the astronomer much prefers to take the one which is easiest to grasp. May be the philosopher demands probability instead; but neither of them will grasp anything certain or hand it on, unless it has been divinely revealed to him Therefore let us, permit these new hypotheses to make a public appearance among old ones which are themselves no more probable, especially since they are wonderful and easy and bring with them a vast storehouse of learn observations. And as far as hypotheses go, let no one expect anything the way of certainty from astronomy, since astronomy can offer us nothing certain, lest, if anyone take as true that which has been-constructed for and other use, he go away from this discipline a bigger fool than when he came to it. Farewell.

PREFACE AND DEDICATION TO POPE PAUL III

I can reckon easily enough, Most Holy Father, that as soon as certain people learn that in these books of mine which I have written about the revolutions of the spheres of the world I attribute certain motions to the terrestrial globe, they will immediately shout to have me and my opinion hooted off stage. For my own works do not please me so much, that I do not weigh what judgments others will pronounce concerning them. And although I realize the conceptions of a philosopher are placed beyond the judgment of the crowd because it is his loving duty to seek the truth in all things, in so far as God in granted that to human reason; nevertheless I think we should avoid opinion utterly foreign. And when I considered how absurd this lecture would be held by those who know that the opinion that the Earth rests immovable in the middle of the heavens as had been confirmed by judgments of many ages—if I were to assert to the contrary that the Earth moves. For a long time I was in great difficulty as to whether I should bring to light my commentaries written to demonstrate the Earth's movement (and wondered) whether it would not be better to follow the example of the Pythagoreans certain others, who used to hand down the mysteries of their philosophy not writing but by word of mouth and only to their relatives, and friends.

But my friends made me change my course in spite of my long-continued hesitation and even resistance.

Dialogues Concerning
the Two Chief World Systems

by Galileo Galilei (1632)

GALILEO GALILEI (1564–1642) was born in Pisa and educated at the university on that city. Among scientists he is honored as the founder of the modern experimental method and for his studies on the behavior of falling bodies. To the general public, however, his fame rests on a different set of accomplishments, having to do with his use of the telescope in astronomy. While working in Venice, he learned of the development of the telescope in Holland and sold an improved version of the instrument to the Arsenal of Venice. In 1610, he turned his improved telescope to the skies and founded the modern science of observational astronomy. At the time, the conventional teaching, based on the writings of Aristotle, was that on Earth there was change and "corruption," but in the heavens there was unchanging perfection. With his new telescope, Galileo saw all sorts of things that contradicted this view, from mountains on the moon to sunspots to the moons of Jupiter. Despite previous problems with Church authorities because of his public espousal of the Copernican system, he published his arguments in the book *Dialogues Concerning the Two Chief World Systems* in which he contrasted the evidence for the Copernican system with that of conventional astronomy, which he associated with Aristotle. In this dialogue, the character Simplicio (the name could be translated as "fool") defends the Aristotelian viewpoint, Sagredo is an earnest seeker after truth, and Salvati is basically Galileo himself. The fact that Galileo put the Pope's own arguments into the mouth of Simplicio obviously did not endear him to his enemies, and doubtless helped to bring about his famous trial and his renunciation of his own work.

* * *

TO THE DISCERNING READER

Several years ago there was published in Rome a salutary edict which, in order to obviate the dangerous tendencies of our present age, imposed a seasonable silence upon the Pythagorean opinion that the earth moves. There were those who impudently asserted that this decree had its origin not injudicious inquire, but in passion none too well informed Complaints were to be heard that advisers who were totally unskilled at astronomical observations ought not to clip the wings of reflective intellects by means of rash prohibitions.

Upon hearing such carping insolence, my zeal could not be contained. Being thoroughly informed about that prudent determination, I decided to appear openly in the theater of the world as a witness of the sober truth. I was at that time in Rome; I was not only received by the most eminent prelates of that Court, but had their applause; indeed this decree was not published without some previous notice of it having been given to me. Therefore I propose in the present work to show to foreign nations that as much is understood of this matter in Italy, and particularly in Rome, as transalpine diligence can ever have imagined. Collecting all the reflections that properly concern the Copernican system, I shall make it known that everything was brought before the attention of the Roman censorship, and that there proceed from this clime not only dogmas for the welfare of the soul, but ingenious discoveries for the delight of the mind as well.

To this end I have taken the Copernican side in the discourse, proceeding as with a pure mathematical hypothesis and striving by every article to represent it as superior to supposing the earth motionless–not, indeed absolutely, but as against the arguments of some professed Peripatetics. These men indeed deserve not even that name, for they do not walk about; they are content to adore the shadows, philosophizing not with due circumspection but merely from having memorized a few ill-understood principles.

THE FIRST DAY
INTERLOCUTORS: SALVIATI, SAGRFDO, AND SIIMPLICIO

[The first part of the dialogue deals with geometrical arguments about the importance of circular motion and the place in the universe. In what follows, Galileo attacks the Aristotelian notion that the universe outside the Earth is eternal and unchanging.]

SAGR. Then what has been said up to now will serve to place under consideration which of two general arguments has the more probability. First there is that of Aristotle, who would persuade us that sublunar bodies are by nature generable and corruptible, etc., and are therefore very different in essence from celestial bodies, these being invariant, ingenerable, incorruptible, etc. This argument is deduced from differences of simple motions. Second is that of Salviati, who assumes the integral parts of the world to be disposed in the best order, and as a necessary consequence excludes straight motions for simple natural bodies as being of no use in nature; he takes the earth to be another of the celestial bodies, endowed with all the prerogatives that belong to them. The latter reasoning suits me better up to this point than the other. Therefore let Simplicio be good enough to produce all the specific arguments, experiments, and observations, both physical and astronomical, by which one may be fully persuaded that the earth differs from the celestial bodies, is immovable, and is located in the center of the universe, or anything else that would exclude the earth from being movable like a planet such as Jupiter, or the moon, etc. And you, Salviati, have the kindness to reply step by step.

SIMP. For a beginning, then, here are two powerful demonstrations proving the earth to be very different from celestial bodies. First, bodies that are generable corruptible, alterable, etc., are quite different from those that are ingenerable, incorruptible, inalterable, etc. The earth is generable, corruptible, alterable, etc., while celestial bodies are ingenerable, incorruptible, inalterable, etc. Therefore the earth is very different from the celestial bodies. ...

Sensible experience shows that on earth there are continual generations, corruptions, alterations, etc., the like of which neither our senses nor the traditions or memories of our ancestors have ever detected in heaven; hence heaven is inalterable, etc., and the earth alterable, etc., and therefore different from the heavens.

The second argument I take from a principal and essential property, which is this: whatever body is naturally dark and devoid of light is different from luminous and resplendent bodies; the earth is dark and without light, and celestial bodies are splendid and full of light; therefore, etc. Answer these, so that too great a pile does not accumulate, and then I will add others.

SALV. As to the first, for whose force you appeal to experience, I wish you would tell me precisely what these alterations are that you see on the earth and not in the heavens, and on account of which you call the earth alterable and the heavens not.

SIMP. On earth I continually see herbs, plants, animals generating and decaying; winds, rains, tempests, storms arising; in a word, the appearance of the earth undergoing perpetual change. None of these changes are to be discerned in celestial bodies, whose positions and configurations correspond exactly with everything men remember, without the generation of anything new there or the corruption of anything old.

SALV. But if you have to content yourself with these visible, or rather these seen experiences, you must consider China and America celestial bodies, since you surely have never seen in them these alterations which you see in Italy. Therefore, in your sense, they must be inalterable.

SIMP. Even if I have never seen such alterations in those places with my own senses, there are reliable accounts of them; besides which those counties being of the earth like ours, they must be alterable like this.

SALV. But why have you not observed this, instead of reducing yourself to having to believe the tales of others? Why not see it with your own eyes?

SIMP. Because those countries are far from being exposed to view; they are so distant that our sight could not discover such alterations in them.

SALV. Now see for yourself how you have inadvertently revealed the fallacy of your argument. You say that alterations which may be seen near at hand on earth cannot be seen in America because of the great distance. Well, so much the less could they be seen in the moon, which is many hundreds of times more distant. And if you believe in alterations in Mexico on the basis of news from there, what reports do you have from the moon to convince you that there are no alterations there? From your not seeing alterations in heaven (where if any occurred you would not be able to see them by reason of the distance, and from whence no news is to be had), you cannot deduce that there are none, in the same way as from seeing and recognizing them on earth you correctly deduce that they do exist here.

SIMP. Among the changes that have taken place on earth I can find some so great that if they had occurred on the moon they could yen well have been observed here below. From the oldest records we have it that formerly, at the Straits of Gibraltar, Abila and Calpe were joined together with some lesser mountains which held the ocean in check; but these mountains being separated by some cause, the opening admitted the sea, which flooded in so as to form the Mediterranean. When we consider the immensity of this, and the difference in appearance which must have been made in the water and land seen from afar, there is no doubt that such a change could easily have been seen by anyone then on the moon. Just so would the inhabitants of earth have discovered any such alteration in the moon; yet there is no history of such a thing being seen. Hence there remains no basis for saying that anything in the heavenly bodies is alterable, etc.

SALV. I do not make bold to say that such great changes have taken place in the moon, but neither am I sure that they could not have happened. Such a mutation could be represented to us only by some variation between the lighter and the darker parts of the moon, and I doubt whether we have had observant selenographers on earth who have for any considerable number of years provided us with such exact selenography as would make us reasonably conclude that no such change has come about in the face of the moon. Of the moon's appearance, I find no more exact description than that some say it represents a human face; others, that it is like the muzzle of a lion; still others, that it is Cain with a bundle of thorns on his back. So to say "Heaven is inalterable, because neither in the moon nor in other celestial bodies are such alterations seen as are discovered upon the earth" has no power to prove anything.

SAGR. This first argument of Simplicio's leaves me with another haunting doubt which I should like to have removed. Accordingly I ask him whether the earth was generable and corruptible before the Mediterranean inundation, or whether it began to be so then?

SIMP. It was without doubt generable and corruptible before, as well; but that was so vast a mutation that it might have been observed as far as the moon.

SAGR. Well, now; if the earth was generable and corruptible before that flood, why may not the moon be equally so without any such change? Why is something necessary in the moon which means nothing on the earth?

SALV. A very penetrating remark. But I am afraid that Simplicio is altering the meaning a bit in this text of Aristotle and the other Peripatetics. They say that they hold the heavens to be inalterable because not one star there has ever been seen to be generated or corrupted, such being probably a lesser part of heaven than a city is of the earth; yet innumerable of the latter have been destroyed so that not a trace of them remains. …

SALV. Now, getting back to the subject, I say that things which are being and have been discovered in the heavens in our own time are such that they can give entire satisfaction to all philosophers, because just such events as we have been calling generations and corruptions have been seen and are being seen in particular bodies and in the whole expanse of heaven. Excellent astronomers have observed many comets generated and dissipated in places above the lunar orbit, besides the two new stars of 1572 and 1604, which were indisputably beyond all the planets. And on the face of the sun itself, with the aid of the telescope, they have seen produced and dissolved dense and dark matter, appearing much like the clouds upon the earth: and many of these are so vast as to exceed not only the Mediterranean Sea, but all of Africa, with Asia thrown in. Now, if Aristotle had seen these things, what do you think he would have said and done, Simplicio?

SIMP. I do not know what would have been done or said by Aristotle, who was the master of all science, but I know to some extent what his followers do and say, and what they ought to do and say in order not to remain without a guide, a leader, and a chief in philosophy.

As to the comets, have not these modern astronomers who wanted to make them celestial been vanquished by the *Anti-Tycho*? Vanquished, moreover, by their own weapons; that is, by means of parallaxes and of calculations turned about every which way, and finally concluding in favor of Aristotle that they are all elemental. A thing so fundamental to the innovators having been destroyed, what more remains to keep them on their feet?

SALV. Calm yourself, Simplicio. What does this modern author of yours say about the new stars of 1572 and 1604, and of the solar spots? As far as the comets are concerned I, for my part, care little whether they are generated below or above the moon, nor have I ever set much store by Tycho's verbosity. Neither do I feel any reluctance to believe that their matter is elemental, and that they may rise as they please without encountering any obstacle from the impenetrability of the Peripatetic heavens, which I hold to be far more tenuous, yielding, and subtle than our air. And as to the calculation of parallaxes, in the first place I doubt whether comets are subject to parallax; besides, the inconstancy of the observations upon which they have been computed renders me equally suspicious of both his opinions and his adversary's—the more so because it seems to me that the *Anti-Tycho* sometimes trims to its author's taste those observations which do not suit his purposes, or else declares them to be erroneous.

SIMP. With regard to the new stars, the *Anti-Tycho* thoroughly disposes of them in a few words, saying that such recent new stars are not positively known to be heavenly bodies, and that if its adversaries wish to prove any alterations and generations in the latter, they must show us mutations made in stars which have already been described for a long time and which are celestial objects beyond doubt. And this can never possibly be done.

As to that material which some say is generated and dissolved on the face of the sun, no mention is made of it at all, from which I should gather that the author takes it for a fable, or for an illusion of the telescope, [note: The telescope was an object of suspicion in many circles.] or at best for some phenomenon produced by the air; in a word, for anything but celestial matter.

SALV. But you, Simplicio, what have you thought of to reply to the opposition of these importunate spots which have come to disturb the heavens, and worse still, the Peripatetic philosophy? It must be that you, as its intrepid defender, have found a reply and a solution which you should not deprive us of.

SIMP. I have heard different opinions on this matter. Some say, "They are stars which, like Venus and Mercury, go about the sun in their proper orbits, and in passing under it present themselves to us as dark; and because there are many of them, they frequently happen to collect together, and then again to separate." Others believe them to be figments of the air; still others, illusions of the lenses; and still others, other things. But I am most inclined to believe—yes, I think it certain—that they are a collection of various different opaque objects, coming together almost accidentally; and therefore we often see that in one spot there can be counted ten or more such tiny bodies of irregular shape that look like snowflakes, or tufts of wool, or flying moths. They change places with each other, now separating and now congregating, but mostly right under the sun, about which, as their center, they move. But it is not therefore necessary to say that they are generated or decay. Rather, they are sometimes hidden behind the body of the sun; at other times, though far from it, they cannot be seen because of their proximity to its immeasurable light.

SALV. If what we are discussing were a point of law or of the humanities, in which neither true nor false exists, one might trust in subtlety of mind and readiness of tongue and in the greater experience of the writers, and expect him who excelled in those things to make his reasoning most plausible, and one might judge it to be the best. But in the natural sciences, whose conclusions are true and necessary and have nothing to do with human will, one must take care not to place oneself in the defense of error; for here a thousand Demostheneses and a thousand Aristotles would be left in the lurch by every mediocre wit who happened to hit upon the truth for himself. Therefore, Simplicio, give up this idea and this hope of yours that there may be men so much more leaned, erudite, and well-read than the rest of us as to he able to make that which is false become true in defiance of nature. And since among all opinions that have thus far been produced regarding the essence of sunspots, this one you have just explained appears to you to be the correct one, it follows that all the rest are false. Now to free you also from that one—which is an utterly delusive chimera—I shall, disregarding the many improbabilities in it, convey to you but two observed facts against it.

One is that many of these spots are seen to originate in the middle of the solar disc, and likewise many dissolve and vanish far from the edge of the sun, a necessary argument that they must be generated and dissolved. For without generation and corruption, they could appear there only by way of local motion, and they all ought to enter and leave by the very edge.

The other observation, for those not in the rankest ignorance of perspective, is that from the changes of shape observed in the spots, and from their apparent changes in velocity, one must infer that the spots are in contact with the sun's body, and that, touching its surface, they are moved either with it or upon it and in no sense revolve in circles distant from it.

Their motion proves this by appearing to be very slow around the edge of the solar disc, and quite fast toward its center; the shapes of the spots prove the same by appearing very narrow around the sun's edge in comparison with how they look in the vicinity of the center. For around the center they are seen in their majesty and as they really are; but around the edge, because of the curvature of the spherical surface, they show themselves foreshortened. These diminutions of both motion and shape, for anyone who knows how to observe them and calculate diligently, correspond exactly to what ought to appear if the spots are contiguous to the sun, and hopelessly contradict their moving in distant circles, or even at small intervals from the solar body. This has been abundantly demonstrated by our mutual friend in his *Letters to Mark Welser on the Solar Spots*. It may be inferred from the same changes of shape that none of these are stars or other spherical bodies, because of all shapes only the sphere is never seen foreshortened, nor can it appear to be anything but perfectly round. So if any of the individual spots were a round body, as all stars are deemed to be, it would present the same roundness in the middle of the sun's disc as at the extreme edge, whereas they so much foreshorten and look so thin near that extremity, and on the other hand so broad and long toward the center, as to make it certain that these are flakes of little thickness or depth with respect to their length and breadth.

Then as to its being observed ultimately that the same spots are sure to return after a certain period, do not believe that, Simplicio; those who said that were trying to deceive you. That this is so, you may see from their having said nothing to you about those that are generated or dissolved on the face of the sun far from the edge; nor told you a word about those which foreshorten, this being a necessary proof of their contiguity to the sun. The truth about the same spots returning is merely what is written in the said *Letters*; namely, that some of them are occasionally of such long duration that they do not disappear in a single revolution around the sun, which takes place in less than a month.

SIMP. To tell the truth, I have not made such long and careful observations that I can qualify as an authority on the facts of this matter; but certainly I wish to do so, and then to see whether I can once more succeed in reconciling what experience presents to us with what Aristotle teaches. For obviously two truths cannot contradict one another.

SALV. Whenever you wish to reconcile what your senses show you with the soundest teachings of Aristotle, you will have no trouble at all. Does not Aristotle say that because of the great distance, celestial matters cannot be treated very definitely?

SIMP. He does say so, quite clearly.

SALV. Does he not also declare that what sensible experience shows ought to be preferred over any argument, even one that seems to be extremely well founded? And does he not say this positively and without a bit of hesitation?

SIMP. He does.

SALV. Then of the two propositions, both of them Aristotelian doctrines, the second—which says it is necessary to prefer the senses over arguments—is a more solid and definite doctrine than the other, which holds the heavens to be inalterable. Therefore it is better Aristotelian philosophy to say "Heaven is alterable because my senses tell me so," than to say, "Heaven is inalterable because Aristotle was so persuaded by reasoning. Add to this that we possess a better basis for reasoning about celestial things than Aristotle did. He admitted such perceptions to be very difficult for him by reason of the distance from his senses, and conceded that one whose senses could better represent them would be able to philosophize about them with more certainty. Now we, thanks to the telescope, have brought the heavens thirty or forty times closer to us than they were to Aristotle, so that we can discern many things in them that he could not see; among other things these sunspots, which were absolutely invisible to him. Therefore we can treat of the heavens and the sun more confidently than Aristotle could.

SAGR. I can put myself in Simplicio's place and see that he is deeply moved by the overwhelming force of these conclusive arguments.

Mathematical Principles of Natural Philosophy

by Isaac Newton (1687)

ISAAC NEWTON (1642–1727) is considered by many to be the greatest scientist who ever lived. Born at his family estate in northern England, he attended Cambridge University. He lived at a time when it was still possible for one man to know all of science, and he made important contributions to many fields, from optics to mathematics. Arguably his most important contribution was the development of the laws of motion that, together with his discovery of the calculus and the law of universal gravitation, allowed scientists to make detailed predictions of the future state of a system if they knew the current state. This led to a picture of the universe in which the planets were like the hands of a clock, while the invisible gears that ran the system could be known through human reason. One thing to note about the following excerpt from *Principia Mathematica* is the formal nature of the writing. This was a legacy of Euclid's *Elements*, which laid out the results of plane geometry in a similar fashion and was routinely studied throughout Europe. Newton's adoption of this style of presentation still influences the way that scientists write about their work today.

* * *

Definition I

The quantity of matter is the measure of the same, arising from its density and bulk conjunctly. Thus air of a double density in a double space is quadruple in quantity; in a triple space, sextuple in quantity. The same thing is to be understood of snow and fine dust or powders, that are condensed by compression or liquefaction; and of all bodies that are by any cause whatever differently condensed. I have no regard in this place to a medium, if any such there is, that freely pervades the interstices between the parts of bodies. It is this quantity that I mean hereafter everywhere under the name of body or mass. And the same is known by the weight of each body; for it is proportional to the weighty as I have found by experiments on pendulums, very accurately made, which shall be shewn hereafter.

Definition II

The quantity of motion is the measure of the same, arising from the velocity and quantity of matter conjunctly.

The motion of the whole is the sum of the motions of all the parts; and therefore in a body double in quantity, with equal velocity, the motion is double; with twice the velocity, it is quadruple.

Definition III

The *vis insita*, or innate force of matter, is a power of resisting, by which every body, as much as in it lies, endeavours to persevere in its present state whether it be of rest or of moving uniformly forward in a right line.

This force is ever proportional to the body whose force it is; and differs nothing from the inactivity of the mass, but in our manner of conceiving it. A body, from the inactivity of matter, is not without difficulty put out of its state of rest or motion. Upon which account, this *vis insita*, may, by a most significant name, be called *vis inertiae*, or force of inactivity. But a body exerts this force only, when another force impressed upon it, endeavours to change its condition; and the exercise of this force may be considered both as resistance and impulse; it is resistance, in so far as the body, for maintaining its present state, withstands the force impressed; it is impulse, in so far as the body, by not easily giving way to the impressed force of another, endeavours to change the state of that other. Resistance is usually ascribed to bodies at rest, and impulse to those in motion; but motion and rest, as commonly conceived, are only relatively distinguished; nor are those bodies always truly at rest, which commonly are taken to be so.

Definition IV

An impressed force is an action exerted upon a body, in order to change its state, either of rest, or of moving uniformly forward in a right line.

This force consists in the action only; and remains no longer in the body, when the action is over. For a body maintains every new state it acquires, by its *vis inertiae* only. Impressed forces are of different origins; as from percussion, from pressure, from centripetal force.

Definition V

A centripetal force is that by which bodies are drawn or impelled, or any way tend, towards a point as to a centre.

Definition VI

The absolute quantity of a centripetal force is the measure of the same, proportional to the efficacy of the cause that propagates it from the centre, through the spaces round about.

Definition VII

The accelerative quantity of a centripetal force is the measure of the same, proportional to the velocity which it generates in a given time.

Definition VIII

The motive quantity of a centripetal force is the measure of the same, proportional to the motion which it generates in a given time.

Thus the weight is greater in a greater body, less in a less body; and, in the same body, it is greater near to the earth, and less at remoter distances. This sort of quantity is the centripetency, or propension of the whole body towards the centre, or, as I may say, its weight; and it is always known by the quantity of an equal and contrary force just sufficient to hinder the descent of the body.

SCHOLIUM

Hitherto I have laid down the definitions of such words as are less known, and explained the sense in which I would have them to be understood in the following discourse. I do not define time, space, place and motion, as being well known to all. Only I must observe, that the vulgar conceive those quantities under no other notions but from the relation they bear to sensible objects. And thence arise certain prejudices, for the removing of which, it will be convenient to distinguish them into absolute and relative, true and apparent, mathematical and common.

I. Absolute, true, and mathematical time, of itself, and from its own nature flows equably without regard to anything external.

II. Absolute space, in its own nature, without regard to anything external, remains always similar and immovable.

III. Place is a part of space which a body takes up, and is according to the space, either absolute or relative.

IV. Absolute motion is the translation of a body from one absolute place into another; and relative motion, the translation from one relative place into another. Thus in a ship under sail, the relative place of a body is that part of the ship which the body possesses; or that part of its cavity which the body fills, and which therefore moves together with the ship: and relative rest is the continuance of the body in the same part of the ship, or of its cavity. But real, absolute rest, is the continuance of the body in the same part of that immovable space, in which the ship itself, and all that it contains, is moved. Wherefore, if the earth is really at rest, the body, which relatively rests in the ship, will really and absolutely move with the same velocity which the ship has on the earth. But if the earth also moves, the true and absolute motion of the body will arise, partly from the true motion of the earth, in immovable space; partly from the relative motion of the ship on the earth; and if the body moves also relatively in the ship; its true motion will arise, partly from the true motion of the earth, in immovable space, and partly from the relative motions as well of the ship on the earth, as of the body in the ship; and from these relative motions will arise the relative motion of the body on the earth. As if that part of the earth, where the ship is, was truly moved toward the east, with a velocity of 10010 parts; while the ship itself, with a fresh gale, and full sails, is carried towards the west, with a velocity expressed by 10 of those parts; but a sailor walks in the ship towards the east, with 1 part of the said velocity; then the sailor will be moved truly in immovable space towards the east, with a velocity of 10001 parts, and relatively on the earth towards the west, with a velocity of 9 of those parts

AXIOMS, OR LAWS OF MOTION
Law I

Every body perseveres in its state of rest, or of uniform motion in a right line, unless it is compelled to change that state by forces impressed thereon.

Projectiles persevere in their motions, so far as they are not retarded by the resistance of the air, or impelled downwards by the force of gravity. A top, whose parts by their cohesion are perpetually drawn aside from rectilinear motions, does not cease its rotation, otherwise than as it is retarded by the air. The greater bodies of the planets and comets, meeting with less resistance in more free spaces, preserve their motions both progressive and circular for a much longer time.

Law II

The alteration of motion is ever proportional to the motive force impressed; and is made in the direction of the right line in which that force is impressed.

If any force generates a motion, a double force will generate double the motion, a triple force triple the motion, whether that force be impressed altogether and at once, or gradually and successively. And this motion (being always directed the same way with the generating force), if the body moved before, is added to or subducted from the former motion, according as they directly conspire with or are directly contrary to each other; or obliquely joined, when they are oblique, so as to produce a new motion compounded from the determination of both.

Law III

To every action there is always opposed an equal reaction: or the mutual actions of two bodies upon each other are always equal, and directed to contrary parts.

Whatever draws or presses another is as much drawn or pressed by that other. If you press a stone with your finger, the finger is also pressed by the stone. If a horse draws a stone tied to a rope, the horse (if I may so say) will be equally drawn back towards the stone: for the distended rope, by the same endeavour to relax or unbend itself, will draw the horse as much towards the stone, as it does the stone towards the horse, and will obstruct the progress' of the one as much as it advances that of the other. If a body impinge upon another, and by its force change the motion of the other, that body also (because of the equality of the mutual pressure) will undergo an equal change, in its own motion, towards the contrary part. The changes made by these actions are equal, not in the velocities but in the motions of bodies; that is to say, if the bodies are not hindered by any other impediments. For, because the motions are equally changed, the changes of the velocities made towards contrary parts are reciprocally proportional.

A Discussion of Elliptical Orbits of Comets
by Edmond Halley (1715)

EDMOND HALLEY (1656–1743), was born near London and educated at Oxford university. He became a prominent astronomer at an early age when he produced the first detailed map of the skies of the southern hemisphere. He participated in many important scientific advances of his day, creating, for example, the first map of magnetic deviations in the Atlantic. He was eventually appointed Astronomer Royal. A friend of Isaac Newton, he used Newton's calculation of comet orbits to predict the return of the comet that now bears his name.

* * *

Hitherto I have consider'd the Orbits of Comets as exactly Parabolic; upon which supposition it wou'd follow, that Comet: being impell'd towards the Sun by a Centripetal Force, would descend as from spaces infinitely distant, and by their so falling acquire such a Velocity, as that they may again fly off into the remotest parts of the Universe, moving upwards with a perpetual tendency, so as never to return again to the Sun. But since they appear frequently enough, and since some of them can be found to move with a Hyperbolic Motion, or a Motion swifter than what a Comet might acquire by its Gravity to the Sun, 'tis highly probable they rather move in very Excentric Elliptic Orbits, and make their returns after long periods of Time: For so their number will be determinate, and, perhaps, not so very great. Besides, the space between the Sun and the Fix'd Stars is so immense, that there is room enough for a Comet to revolve, tho' the Period of its revolution be vastly long.

The principal use therefore, of this Table of the Elements of their Motions [i.e. the numbers that determine the shape of the elliptical orbits of the comets], and that which indeed induced me to construct it, is, that whenever a new Comet shall appear, we may be able to know, by comparing together the Elements, whether it be any of those which has appear'd before, and consequently to determine its Period, and the Axis of its Orbit, and to foretel its Return. And, indeed there are many things which make me believe that the Comet which *Apian* observed in the Year 1531, was the same with that which *Kepler* and *Longomontanus* more accurately describ'd in the Year 1607: and which I myself have seen return, and observ'd in the Year 1682. All the Elements agree, and nothing seems to contradict this my opinion, besides the Inequality of the Periodic revolutions. Which Inequality is not so great neither as that it may not be owing to Physical Causes, For the Motion of Saturn is so disturbed by the rest of the Planets, especially Jupiter, that the Periodic time of that Planet is uncertain for some whole days together. How much more therefore will a Comet be subject to such like errors, which rises almost four times higher than Saturn, and whose Velocity, tho' increased but a very little, would be sufficient to change its Orbit, from an Elliptical to a Parabolical one. And I am the more confirmed in my opinion of its being the same; for that in the Year 1456, in the Summer time, a Comet was seen passing Retrograde between the Earth and the Sun, much after the same manner: Which tho' nobody made observations upon it, yet from its Period and the manner of its Transit, I cannot think different from those I have just now mention'd. And since looking over the Histories of Comets I find, at an equal interval of Time, a Comet to have been seen about Easter in the Year 1305, which is another double Period of 151 Years before the former. Hence I think I may venture to foretel, that it will return again in the Year 1758. And, if it should then so return, we shall have no reason to doubt but the rest may return also: Therefore, Astronomers have a large field wherein to exercise themselves for many ages, before they will be able to know the number of these many and great Bodies revolving about the common Center of the Sun, and to reduce their Motions to certain Rules.

CHAPTER 3
ENERGY

The many different forms of energy are interchangeable, and the total amount of energy in an isolated system is conserved.

INTRODUCTION

THE FIRST LAW of thermodynamics states that the many forms of energy are interchangeable, while the total amount of energy in a closed system is unchanged. The following two excerpts explore crucial aspects of the first law.

Benjamin Thompson, later Count Rumford, (1753–1814) was born in Massachusetts. He sided with the British cause in the Revolutionary War and fled to London after the conflict, joining the British Army and serving as undersecretary for the colonies in the British government. After retiring he moved to Bavaria, where he became head of the Bavarian army (this is when he became "Count Rumford"). His observations of cannon manufacturing led to the conclusion that heat was not a fluid that moved between objects, but a manifestation of what would be called atomic motion today. Heat is thus one of the many forms of energy.

An Inquiry Into the Source of Heat Which is Excited by Friction

by Benjamin Thompson, Count Rumford (1798)

IT FREQUENTLY HAPPENS that in the ordinary affairs and occupations of life opportunities present themselves of contemplation some of the most curious operations of nature; and very interesting philosophical experiments might often be made, almost without trouble or expense, by means of machinery contrived for the mere mechanical purposes of the arts and manufactures.

I have frequently had occasion to make this observation, and am persuaded that a habit of keeping the eyes open to everything that is going on in the ordinary course of the business of life has oftener led, as it were by accident, or in the playful excursions of the imagination, put into action by contemplating the most common appearance, than all the more intense meditation of philosophers, in the hours expressly set apart for study.

It was by accident that I was led to make the experiments of which I am to give an account; and, though they are not perhaps of sufficient importance to merit so formal an introduction, I cannot help flattering myself that they will be thought curious in several respects, and worthy of the honor of being made known to the Royal Society.

Being engaged, lately, in superintending the boring of cannon, in the workshops of the military arsenal at Munich, I was struck with the very considerable degree of heat which a brass gun acquires, in a short time, in being bored; and with the still more intense heat (much greater than that of boiling water, as I found by experiment) of the metallic chips separated from it by the borer.

The more I meditated on these phenomena the more they appeared to me to be curious and interesting. A thorough investigating of them seemed even to bid fair to give a farther insight into the hidden nature of heat; and to enable us to form some reasonable conjectures respecting the existence, or non-existence, of an igneous fluid: a subject on which the opinions of philosophers have, in all ages, been much divided.

In order that the society may have clear and distinct ideas of the speculations and also of the specific objects of philosophical investigation they suggested to me, I must beg leave to state them at some length and in such manner as I shall think best to answer this purpose.

From whence comes the heat actually produced in the mechanical operation above mentioned?

Is it furnished by the metallic chips which are separated by the borer from the solid mass of metal?

If this were the case, then, according to the modern doctrines of latent heat, and of caloric, the capacity for the heat of the parts of the metal, so reduced to chips, ought not only to be changed, but the change undergone by them should be sufficiently great to account for all the heat produced.

But no such change had taken place; for I found, upon taking equal quantities, by weight, of these chips, and of thin slips of the same block of metal separated by means of a fine saw, and putting them at the same temperature (that of boiling water) into equal quantities of cold water (that is to say, at the temperature of 59 1/2 °F), the portion of the water into which the chips were put was not, to all appearance, heated either less or more than the other portion, in which the slips of metal were to put.

This experiment being repeated several times, the results were always so nearly the same that I could not determine whether any, or what change, had been produced in the metal, in regard to its capacity for heat, by being reduced to chips by the borer.

From hence it is evident that the heat produced could not possibly have been furnished at the expense of the latent heat of the metallic chips. But, not being willing to rest satisfied with these trials, however conclusive they appeared to me to be, I had resource to the following still more decisive experiment:

Taking a cannon (a brass six-pounder) cast solid, and rough as it came from the foundry, and fixing it (horizontally) in the machine used for boring, and at the same time finishing the outside of the cannon by turning, I caused its extremity to be cut off; and, by turning down the metal in that part, a solid cylinder was formed, 7 3/4 inches in diameter, and 9 8/10 inches long.

This short cylinder, which was supported in its horizontal position, and turned round its axis, by means of the neck by which it remained united to the cannon, was now bored with the horizontal borer used in boring cannon.

This cylinder being designed for the express purpose of generating heat by friction, by having a blunt borer forced against its solid bottom at the same time that it should be turned round its axis by the force of horses, in order that the heat accumulated in the cylinder might from time to time be measured, a small round hole, 0.37 of an inch only in diameter, and 4.2 inches in depth, for the purpose of introduction a small cylindrical mercurial thermometer, was made in it.

This experiment was made in order to ascertain how much heat was actually generated by friction, when a blunt steel borer being so forcibly shoved (by means of a strong screw) against the bottom of the bore of the cylinder that the pressure against it was equal to the weight of about 10,000 pounds avoirdupois, the cylinder was turned round on its axis (by the force of horses) at the rate of about thirty-two times in a minute. ...

To prevent, as far as possible, the loss of any part the heat that was generated in the experiment, the cylinder was well covered up with a fit coating of thick and warm flannel, the cylinder was carefully wrapped round it, and defended it on every side from the cold air of the atmosphere.

At the beginning of the experiment the temperature of the air in the shade, as also that of the cylinder, was just 60 °F.

At the end of thirty minutes, when the cylinder had made 960 revolutions about its axis, the horses being stopped, a cylindrical mercurial thermometer, whose bulb was 32/100 of an inch in diameter, and 3 1/4 inches in length, was introduced into the hole made to receive it, in the side of the cylinder, when the mercury rose almost instantly to 130 °F. ...

Finding so much reason to conclude that the heat generated in these experiments, or excited, as I would rather choose to express it, was not furnished at the expense of the latent heat or combined caloric of the metal, I pushed my inquiries a step farther and endeavored to find out whether the air did, or did not, contribute anything in the generation of it. ...

Everything being ready, I proceeded to make the experiment I had projected in the following manner:

The hollow cylinder having been previously cleaned out, and the inside of its bore wiped with a clean towel till it was quite dry, the square iron bar, with the blunt steel borer fixed to the end of it, it was put into its place; the mouth of the bore of the cylinder being closed at the same time, by means of the circular piston, through the center of which the iron bar passed.

This being done, the box was put in its place, and the joining of the iron rod, and of the neck of the cylinder, with the two ends of the box, having been with cold water (viz., at the temperature of 60 °F) and the machine was put in motion.

The result of this beautiful experiment was very striking, and the pleasure it afforded me amply repaid me for all the trouble I had had in contriving and arranging the complicated machinery used in making it.

The cylinder, revolving at the rate of about thirty-two times in a minute, had been in motion but a short time when I perceived, by putting my hand into the water and touching the outside of the cylinder, that heat was generated; and it was not long before the water which surrounded the cylinder began to be sensibly warm.

At the end of one hour I found, by plunging a thermometer into the water in the box (the quantity of which fluid amounted to 18.77 pounds avoirdupois, or 2 1/4 wine gallons) that its temperature had been raised no less than 47 degrees; being now 107° of Fahrenheit's scale.

When thirty minutes more had elapsed, or one hour and thirty minutes after the machinery had been put in motion, the heat of the water in the box was 142 °F.

At the end of two hours, reckoning from the beginning of the experiment, the temperature of the water was found to be raised to 178 °F.

At two hours twenty minutes it was 200 °F; and at two hours thirty minutes it *actually boiled!*

It would be difficult to describe the surprise and astonishment expressed in the countenances of the bystanders, on seeing so large a quantity of cold water heated and actually made to boil without any fire.

Though there was, in fact, nothing that could justly be considered as surprising in this event, yet I acknowledge fairly that it afforded me a degree of childish pleasure, which, were I ambitious of the reputation of a grave philosopher, I ought most certainly rather to hide than to discover.

The quantity of heat excited and accumulated in this experiment was very considerable; for not only the water in the box, but also the box itself (which weighed 15 1/4 pounds) and the hollow metallic cylinder, and that apart of the iron bar which, being situated within the cavity of the box, was immersed in the water, were heated 150 degrees of Fahrenheit's scale; viz., from 60 °F (which was the temperature of the water, and of the machinery, at the beginning of the experiment) to 210 °F, the heat of boiling water at Munich.

The total quantity of the heat generated may be estimated with some considerable degree of precision. ...

From the result of these computations it appears that the quantity of heat produced equably, or in a continual stream (if I may use that expression), by the friction of the blunt steel borer against the bottom of the hollow metallic cylinder, in the experiment under consideration, was greater than that produced equably in the combustion of nine wax candles, each three quarters of an inch in diameter, all burning together, or at the same time, with clear bright flames.

As the machinery used in this experiment could easily be carried round by the force of one horse (though, to render the work lighter, two horses were actually employed in doing it), these computations show further how large a quantity of heat might be produced by proper mechanical contrivance, merely by the strength of a horse, without either fire, light, combustion, or chemical decomposition; and, in a case of necessity, the heat thus produces might be used in cooking victuals.

But no circumstances can be imagined in which this method of procuring heat would not be disadvantageous; for more heat might be obtained by using the fodder necessary for the support of a horse, as fuel.

By meditating on the results of all these experiments we are naturally brought to that great question which has so often been the subject of speculation among philosophers; namely:

What is heat? Is there any such thing as an *igneous fluid*? Is there anything that can with propriety be called *caloric*?

We have seen that a very considerable quantity of heat may be excited in the friction of two metallic surfaces and given off in a constant stream or flux, *in all directions*, without iteration or intermission, and without any signs of diminution or exhaustion.

From whence came the heat which was continually given off in this manner, in the foregoing experiments? Was it furnished by the small particles of metal, detached from the larger solid masses, on their being rubbed together? This, as we have already seen, could not possibly have been the case.

Was it furnished by the air? This could not have been the case; for in three of the experiments, the machinery being kept immersed in water, the access of the air of the atmosphere was completely prevented.

Was it furnished by the water which surrounded the machinery? That this could not have been the case is evident: first, because this water was continually *receiving heat* from the machinery and could not, at the same time, be *giving to*, and *receiving heat from*, the same body; and secondly, because there was no chemical decomposition of any part of this water. had any such decomposition taken place (which indeed could not reasonably have been expected), one of its component elastic fluids (most probably inflammable air) must, at the same time, have been set at liberty, and in making its escape into the atmosphere would have been detected; but though I frequently examined the water to see if any air bubbles rose up through it, and had even made preparations for catching them, in order to examine them, if any should appear, I could perceive none; nor was there any sign of decomposition of any kind whatever, or other chemical process, going on in the water.

Is it possible that the heat could have been supplied by means of the iron bar to the end of which the blunt steel borer was fixed? Or by the small neck of gun metal by which the hollow cylinder was united to the cannon? These suppositions appear more improbable even than either of these before mentioned; for heat was continually going off, or *out of the machinery*, by both these passages, during the whole time the experiment lasted.

And, in reasoning on this subject, we must not forget to consider that most remarkable circumstance, that the source of the heat generated by friction, in these experiments, appeared evidently to be *inexhaustible*.

It is in hardly necessary to add that anything which any *insulated* body, or system of bodies, can continue to furnish *without limitation* cannot possibly be a *material substance*; and it appears to me to be extremely difficult, if not quite impossible, to form any distinct idea of anything, capable of being excited and communicated, in the manner the heat was excited and communication in these, except it be MOTION.

On the Secular Cooling of the Earth
by William Thomson, Lord Kelvin (1864)

WILLIAM THOMSON (LORD Kelvin) (1824–1907) was born in Ireland and educated at the University of Glasgow. He was one of the giants of nineteenth century science, making major contributions to many fields. In addition to being one of the main architects of the science of thermodynamics, for example, he was one of the directors in the company that laid the first trans-Atlantic telegraph cable. In this paper, however, he initiated a long conflict between physicists and earth scientists over the question of the age of the Earth. Kelvin's argument about the cooling of the Earth was a direct consequence of the conservation of energy. Given the current rate of cooling, he argued, the Earth could not possibly be old enough to allow for the immense times needed for geological formations and complex living organisms to form. His arguments, firmly based on the first law of thermodynamics, were eventually rendered moot by the discovery of radioactivity—a source of energy unknown in Kelvin's time.

* * *

1. For eighteen years it has pressed on my mind, that essential principles of Thermodynamics have been overlooked by those geologists who uncompromisingly oppose all paroxysmal hypostheses, and maintain not only that we have examples now before us, on the earth, of all the different actions by which its crust has been modified in geological history, but that these actions have never, or have not on the whole, been more violent in past time than they are at present.

2. It is quite certain the solar system cannot have gone on even as at present, for a few hundred thousand or a few million years, without the irrevocable loss (by dissipation, not by *annihilation*) of a very considerable proportion of the entire energy initially in store for sun heat, and for Plutonic action. It is quite certain that the whole store of energy in the solar system has been greater in all past time, than at present; but it is conceivable that the rate at which it has been drawn upon and dissipated, whether by solar radiation, or by volcanic action in the earth or other dark bodies of the system, may have been nearly equable, or may even have been less rapid, in certain periods of the past. But it is far more probable that the secular rate of dissipation has been in some direct proportion to the total amount of energy in store, at any time after the commencement of the present order of things, and has been therefore very slowly diminishing from age to age.

3. I have endeavoured to prove this for the sun's heat, in an article recently published in *Macmillan's Magazine*,[1] where I have shown that most probably the sun was sensibly hotter a million years ago than he is now. Hence, geological speculations assuming somewhat greater extremes of heat, more violent storms and floods, more luxuriant vegetation, and hardier and coarser-grained plants and animals, in remote antiquity, are more probable than those of the extreme quietist, or "uniformitarian," school. A "middle path," not generally safest in scientific speculation, seems to be so in this case. It is probable that hypotheses of grand catastrophes destroying all life from the earth, and raining its whole surface at once, are greatly in error; it is impossible that hypotheses assuming an equability of sun and storm for 1,000,000 years, can be wholly true.

4. Fourier's mathematical theory of the conduction of heat is a beautiful working out of a particular case belonging to the general doctrine of the "Dissipation of Energy."[2] A characteristic of the practical solutions it presents is, that in each case a distribution of temperature, becoming gradually equalised through an unlimited future, is expressed as a function of the time, which is infinitely divergent for all times longer past than a definite determinable. ...

5. The chief object of the present communication is to estimate from the known general increase of temperature in the earth downwards, the date of the first establishment of that *consistentior status*, which, according to Leibnitz's theory, is the initial date of all geological history.

6. In all parts of the world in which the earth's crust has been examined, at sufficiently great depths to escape influence of the irregular and of the annual variations of the superficial temperature, a gradually increasing temperature has been found in going deeper. The rate

of augmentation (estimated at only 1/110th of a degree, Fahrenheit., in some localities, and as much as 1/15th of a degree in others, per foot of descent) has not been observed in a sufficient number of places to establish any fair average estimate for the upper crust of the whole earth. But 1/50th is commonly accepted as a rough mean; or, in other words, it is assumed as a result of observation, that there is, on the whole, about 1° Fahr. of elevation of temperature per 50 British feet of descent.

7. The fact that the temperature increases with the depth implies a continual loss of heat from the interior, by conduction outwards through or into the upper crust. Hence, since the upper crust does not become hotter from year to year, there must be a secular loss of heat from the whole earth. It is possible that no cooling may result from this loss of heat, but only an exhaustion of potential energy, which in this case could scarcely be other than chemical affinity between substances forming part of the earth's mass. But it is certain that either the earth is becoming on the whole cooler from age to age, or the heat conducted out is generated in the interior by temporary dynamical (that is, in this case, chemical) action.[5] To suppose, as Lyell, adopting the chemical hypothesis, has done,[6] that the substances, combining together, may be again separated electrolytically by thermo-electric currents, due to the heat generated by their combination, and thus the chemical action and its heat continued in an endless cycle, violates the principles of natural philosophy in exactly the same manner, and to the same degree, as to believe that a clock constructed with a self-winding movement may fulfil the expectations of its ingenious inventor by going for ever.

8. It must indeed be admitted that many geological writers of the Uniformitarian school, who in other respects have taken a profoundly philosophical view of their subject, have argued in a most fallacious manner against hypotheses of violent action in past ages. If they had contented themselves with showing that many existing appearances, although suggestive of extreme violence and sudden change, may have been brought about by long-continued action, or by paroxysms not more intense than some of which we have experience within the periods of human history, their position might have been unassailable; and certainly could not have been assailed except by a detailed discussion of their facts. It would be a very wonderful, but not an absolutely incredible result, that volcanic action has never been more violent on the whole than during the last two or three centuries; but it is as certain that there is now less volcanic energy in the whole earth than there was a thousand years ago, as it is that there is less gunpowder in the "Monitor" after she has been seen to discharge shot and shell, whether at a nearly equable rate or not, for five hours without receiving fresh supplies, than there was at the beginning of the action. Yet this truth has been ignored or denied by many of the leading geologists of the present day,[7] because they believe that the facts within their province do not demonstrate greater violence in ancient changes of the earth's surface, or do demonstrate a nearly equable action in all periods.

9. The chemical hypothesis to account for underground heat might be regarded as not improbable, if it was only in isolated localities that the temperature was found to increase

with the depth; and, indeed, it can scarcely be doubted that chemical action exercises an appreciable influence (possibly negative, however) on the action of volcanoes; but that there is slow uniform "combustion," *eremacausis*, or chemical combination of any kind going on, at some great unknown depth under the surface everywhere, and creeping inwards gradually as the chemical affinities in layer after layer are successively saturated, seems extremely improbable, although it cannot be pronounced to be absolutely impossible, or contrary to all analogies in nature. The less hypothetical view, however, that the earth is merely a warm chemically inert body cooling, is clearly to be preferred in the present state of science.

10. Poisson's celebrated hypothesis, that the present underground heat is due to a passage, at some former period, of the solar system through hotter stellar regions, cannot provide the circumstances required for a palaeontology continuous through that epoch of external heat. For from a mean of values of the conductivity, in terms of the thermal capacity of unit volume, of the earth's crust, in three different localities near Edinburgh, which I have deduced from the observations on underground temperature instituted by Principal Forbes there, I find that if the supposed transit through a hotter region of space tool; place between 1250 and 5000 years ago, the temperature of that supposed region must have been from 25° to 50° Fahr. above the present mean temperature of the earth's surface, to account for the present general rate of underground increase of temperature, taken as 1° Fahr. in 50 feet downwards. Human history negatives this supposition. Again, geologists and astronomers will, I presume, admit that the earth cannot, 20,000 years ago, have been in a region of space 100° Fahr. warmer than its present surface. But if the transition from a hot region to a cool region supposed by Poisson took place more than 20,000 years ago, the excess of temperature must have been more than 100° Fahr., and must therefore have destroyed animal and vegetable life. Hence, the farther back and the hotter we can suppose Poisson's hot region, the better for the geologists who require the longest periods; but the best for their view is Leibnitz's theory, which simply supposes the earth to have been at one time an incandescent liquid, without explaining how it got into that state. If we suppose the temperature of melting rock to be about 10,000° Fahr. (an extremely high estimate), the consolidation may have taken place 200,000,000 years ago. Or, if we suppose the temperature of melting rock to be 7000° Fahr. (which is more nearly what it is generally assumed to be), we may suppose the consolidation to have taken place 98,000,000 years ago.

11. These estimates are founded on the Fourier solution demonstrated below. The greatest variation we have to make on them, to take into account the differences in the ratios of conductivities to specific heats of the three Edinburgh rocks, is to reduce them to nearly half, or to increase them by rather more than half. A reduction of the Greenwich underground observations recently communicated to me by Professor Everett of Windsor, Nova Scotia [now, 1889, of Queen's College, Belfast], gives for the Greenwich rocks a quality intermediate between those of the Edinburgh rocks. But we are very ignorant as to the effects of high temperatures in altering the conductivities and specific heats of rocks, and as to their latent

heat of fusions. We must, therefore, allow very wide limits in such an estimate as I have attempted to make; but I think we may with much probability say that the consolidation cannot have taken place less than 20,000,000 years ago, or we should have more underground heat than we actually have, nor more than 400,000,000 years ago, or we should not have so much as the least observed underground increment of temperature. That is to say, I conclude that Leibnitz's epoch of emergence of the *consistentior status* was probably between those dates.

12. The mathematical theory on which these estimates are founded is very simple, being in fact merely an application of one of Fourier's elementary solutions to the problem of finding at any time the rate of variation of temperature from point to point, and the actual temperature at any point, in a solid extending to infinity in all directions, on the supposition that at an initial epoch the temperature has had two different constant values on the two sides of a certain infinite plane. ...

24. How the temperature of solidification, for any pressure, may be related to the corresponding temperature of fluid convective equilibrium, it is impossible to say, without knowledge, which we do not yet possess, regarding the expansion with heat, and the specific heat of the fluid, and the change of volume, and the latent heat developed in the transition from fluid to solid.

25. For instance, supposing, as is most probably true, both that the liquid contracts in cooling towards its freezing-point, and that it contracts in freezing, we cannot tell, without definite numerical data regarding those elements, whether the elevation of the temperature of solidification, or of the actual temperature of a portion of the fluid given just above its freezing-point, produced by a given application of pressure, is the greater. If the former is greater than the latter, solidification would commence at the bottom, or at the center, if there is no solid nucleus to begin with, and would proceed outwards, and there could be no complete permanent incrustation all round the surface till the whole globe is solid, with, possibly, the exception of irregular, comparatively small spaces of liquid.

26. If, on the contrary, the elevation of temperature, produced by an application of pressure to a given portion of the fluid, is greater than the elevation of the freezing temperature produced by the same amount of pressure, the superficial layer of the fluid would be the first to reach its freezing-point, and the first actually to freeze.

27. But if, according to the second supposition of §22 above, the liquid expanded in cooling near its freezing point, the solid would probably likewise be of less specific gravity than the liquid at its freezing-point. Hence the surface would crust over permanently with a crust of solid, constantly increasing inwards by the freezing of the interior fluid in consequence of heat conducted out through the crust. The condition most commonly assumed by geologists would thus be produced.

28. But Bischof's experiments, upon the validity of which, so far as I am aware, no doubt has ever been thrown, show that melted granite, slate, and trachyte, all contracted by

something about 20 per cent. in freezing. We ought, indeed, to have more experiments on this most important point, both to verify Bischof's results on rocks, and to learn how the case is with iron and other unoxydised metals. In the meantime we must assume it as probable that the melted substance of the earth did really contract by a very considerable amount in becoming solid.

29. Hence, if according to any relations whatever among the complicated physical circumstances concerned, freezing did really commence at the surface, either all round or in any part, before the whole globe had become solid, the sollidified superficial layer must have broken up and sunk to the bottom, or to the centre, before it could have attained a sufficient thickness to rest stably on the lighter liquid below. It is quite clear, indeed, that if at any time the earth were in the condition of a thin solid shell of, let us suppose 50 feet or 100 feet thick of granite, enclosing a continuous melted mass of 20 per cent. less specific gravity in its upper parts, where the pressure is small, this condition cannot have lasted many minutes. The rigidity of a solid shell of superficial extent, so vast in comparison with its thickness, must be as nothing, and the slightest disturbance must cause some part to bend down, crack, and allow the liquid to run out over the whole solid. The crust itself must in consequence become shattered into fragments, which must all sink to the bottom, or to meet in the centre and form a nucleus there if there is none to begin with.

30. It is, however, scarcely possible, that any such continuous crust can ever have formed all over the melted surface at one time, and afterwards have fallen in. The mode of solidification conjectured in §25, seems on the whole the most consistent with what we know of the physical properties of the matter concerned. So far as regards the result, it agrees, I believe, with the view adopted as the most probable by Mr. Hopkins. But whether from the condition being rather that described in §26, which seems also possible, for the whole or for some parts of the heterogeneous substance of the earth, or from the viscidity as of mortar, which necessarily supervenes in a melted fluid, composed of ingredients becoming, as the whole cools, separated by crystallising at different temperatures before the solidification is perfect, and which we actually see in lava from modern volcanoes; it is probable that when the whole globe, or some very thick superficial layer of it, still liquid or viscid, has cooled down to near its temperature of perfect solidification, incrustation at the surface must commence.

31. It is probable that crust may thus form over wide extents of surface, and may be temporarily buoyed up by the vesicular character. It may have retained from the ebullition of the liquid in some places, or, at all events, it may be held up by the viscidity of the liquid; until it has acquired some considerable thickness sufficient to allow gravity to manifest its claim, and sink the heavier solid below the lighter liquid. This process must go on until the sunk portions of crust build up from the bottom a sufficiently close ribbed solid skeleton or frame, to allow fresh incrustations to remain bridging across the now small areas of lava pools or lakes.

32. In the honey-combed solid and liquid mass thus formed, there must be a continual tendency for the liquid, in consequence of its less specific gravity, to work its way up; whether by masses of solid falling from the roof of vesicles or tunnels, and causing earthquake shocks, or by the roof breaking quite through when very thin, so as to cause two such hollows to unite, or the liquid of any of them to flow out freely over the outer surface of the earth; or by gradual subsidence of the solid, owing to the thermodynamic melting, which portions of it, under intense stress, must experience, according to views recently published by my brother, Professor James Thomson. The results which must follow from this tendency seem sufficiently great and various to account for all that we see at present, and all that we learn from geological investigation, of earthquakes, of upheavals and subsidences of solid, and of eruptions of melted rock.

33. These conclusions, drawn solely from a consideration of the necessary order of cooling and consolidation, according to Bischof's result as to the relative specific gravities of solid and of melted rock, are in perfect accordance with what I have recently demonstrated regarding the present condition of the earth's interior,—that it is not, as commonly supposed, all liquid within a thin solid crust of from 30 to 100 miles thick, but that it is on the whole more rigid certainly than a continuous solid globe of glass of the same diameter, and probably than one of steel.

CHAPTER 4
HEAT AND THE SECOND LAW OF THERMODYNAMICS

Heat is a form of energy that flows from warmer to cooler objects.

INTRODUCTION

U NLIKE NEWTON'S LAWS of motion, no one individual discovered the laws of thermodynamics. The understanding of heat and energy developed gradually, through studies of both abstract mathematical concepts such as efficiency and entropy and very practical devices like steam engines and thermometers. The following two excerpts were just such practical contributions.

Daniel Fahrenheit (1686–1736) was born in Poland and moved to Amsterdam at the age of 15, where he became a scientific instrument maker. As part of his work manufacturing thermometers, he developed the standard temperature scale that now bears his name. Notice that he took his high temperature standard to be "blood temperature" and (mistakenly) set it to be 100 degrees. As thermometers improved, it became clear that human body temperature is actually 98.6 degrees on Fahrenheit's original scale.

Experiments Done on the Degree of Heat of a Few Boiling Liquids[1]

by Daniel Fahrenheit (1724)

ABOUT TEN YEARS ago I read in the History of the Sciences issued by the Royal Academy of Paris, that the celebrated Amontons, using a thermometer of his own invention, had discovered that water boils at a fixed degree of heat. I was at once inflamed with a great desire to

---

1   Latin translation by William Francis Magie from *Experimenta Circa Gradum Caloris Liquorum Nonnullorum Ebullientium Instituta.*

make for myself a thermometer of the same sort, so that I might with my own eyes perceive this beautiful phenomenon of nature, and be convinced of the truth of the experiment.

I therefore attempted to construct a thermometer, but because of my lack of experience in its construction, my efforts were in vain, though they were often repeated; and since other matters prevented my going on with the development of the thermometer, I postponed any further repetition of my attempts to some more fitting time. Though my powers and my time failed me, yet my zeal did not slacken, and I was always desirous of seeing the outcome of the experiment. It then came into my mind what that most careful observer of natural phenomena had written about the correction of the barometer; for he had observed that the height of the column of mercury in the barometer was a little (though sensibly enough) altered by the varying temperature of the mercury. From this I gathered that a thermometer might perhaps be constructed with mercury, which would not be so hard to construct, and by the use of which it might be possible to carry out the experiment which I so greatly desired to try. When a thermometer of that sort was made (perhaps imperfect in many ways) the result answered to my prayer; and with great pleasure of mind I observed the truth of the thing.

Three years then passed, in which I was occupied with optical and other work, when I became anxious to try by experiment whether other liquids boiled at fixed degrees of heat.

The results of my experiments are contained in the following table, of which the first column contains the liquids used, the second their specific gravity, the third, the degree of heat which each liquid attains when boiling.

Liquids	Specific Gravity of Liquids at 48° of Heat	Degree Attained by Boiling
Spirits of Wine or Alcohol	8260	176
Rain Water	10000	212
Spirits of Niter	12935	142
Lye Prepared from Wine Lees	<15634	240
Oil of Vitriol	18775	546

I thought it best to give the specific gravity of each liquid, so that, if the experiments of others already tried, or which may be tried, give different results, if might be determined whether the difference should be looked for as resulting from differences in the specific gravities or from other causes. The experiments were not made at the same time, and hence the liquids were affected by different degrees of temperature or heat, but since their gravity is altered in a different way and unequally, I reduced it by calculation to the degree 48, which in my thermometers holds the middle place between the limit of the most intense cold obtained artificially in a mixture of water, of ice and of sal-ammoniac or even of sea-salt, and the limit of heat which is found in the blood of a healthy man.

Volatile oils certainly begin to boil at a fixed degree, but their heat is always increased by boiling. Perhaps this is because the more volatile particles readily leave while resinous particles possessed of greater attraction remain. On the other hand fixed oils are affected by so much heat that the Mercury of the thermometer begins to boil at the same time as they do and hence their heat can hardly be explored by the aforementioned method. But I devised another method which I hope to have the honor of describing in another session before the illustrious Royal Society.

 # Reflections on the Motive Power of Fire and on Machines Fitted to Develop that Power[*]

by Sadi Carnot (1824)

SADI CARNOT (1796–1832) WAS born into a prominent French family who were supporters of Napoleon. He entered the prestigious *Ecole Polytechnique* at the age of 16 and eventually joined the General Staff Corps of the French army. Realizing that the ascendance of England was due in no small part to the development of the steam engine, he decided to develop a solid general theory that explained that device. In this paper he developed the theory of what is now called the Carnot cycle, showing that an engine could work only if it transferred heat from a hot to a cold reservoir. After his untimely death, it was realized that this was an alternate statement of the second law of thermodynamics.

* * *

Every one knows that heat can produce motion. That it possesses vast motive-power no one can doubt, in these days when the steam-engine is everywhere so well known.

To heat also are due the vast movements which take place on the earth. It causes the agitations of the atmosphere, the ascension of clouds, the fall of rain and of meteors, the currents of water which channel the surface of the globe, and of which man has thus far employed but a small portion. Even earthquakes and volcanic eruptions are the result of heat.

From this immense reservoir we may draw the moving force necessary for our purposes. Nature, in providing us with combustibles on all sides, has given us the power to produce, at

---

[*] Translated and edited by R. H. Thurston.

all times and in all places, heat and the impelling power which is the result of it. To develop this power, to appropriate it to our uses, is the object of heat-engines.

The study of these engines is of the greatest interest, their importance is enormous, their use is continually increasing, and they seem destined to produce a great revolution in the civilized world.

Already the steam-engine works our mines, impels our ships, excavates our ports and our rivers, forges iron, fashions wood, grinds grains, spins and weaves our cloths, transports the heaviest burdens, etc. It appears that it must some day serve as a universal motor, and be substituted for animal power, waterfalls, and air currents.

Over the first of these motors it has the advantage of economy, over the two others the inestimable advantage that it can be used at all times and places without interruption.

If, some day, the steam-engine shall be so perfected that it can be set up and supplied with fuel at small cost, it will combine all desirable qualities, and will afford to the industrial arts a range the extent of which can scarcely be predicted. It is not merely that a powerful and convenient motor that can be procured and carried anywhere is substituted for the motors already in use, but that it causes rapid extension in the arts in which it is applied, and can even create entirely new arts.

The most signal service that the steam-engine has rendered to England is undoubtedly the revival of the working of the coal mines, which had declined, and threatened to cease entirely, in consequence of the continually increasing difficulty of drainage, and of raising the coal.[1] We should rank second the benefit to iron manufacture, both by the abundant supply of coal substituted for wood just when the latter had begun to grow scarce, and by the powerful machines of all kinds, the use of which the introduction of the steam-engine has permitted or facilitated.

Iron and heat are, as we know, the supporters, the bases, of the mechanic arts. It is doubtful if there be in England a single industrial establishment of which the existence does not depend on the use of these agents, and which does not freely employ them. To take away today from England her steam-engines would be to take away at the same time her coal and iron. It would be to dry up all her sources of wealth, to ruin all on which her prosperity depends, in short, to annihilate that colossal power. The destruction of her navy, which she considers her strongest defence, would perhaps be less fatal.

The safe and rapid navigation by steamships may be regarded as an entirely new art due to the steam-engine. Already this art has permitted the establishment of prompt and regular communications across the arms of the sea, and on the great rivers of the old and new continents. It has made it possible to traverse savage regions where before we could scarcely penetrate. It has enabled us to carry the fruits of civilization over portions of the globe where they would else have been wanting for years. Steam navigation brings nearer together the most distant nations. It tends to unite the nations of the earth as inhabitants of one country.

In fact, to lessen the time, the fatigues, the uncertainties, and the dangers of travel—is not this the same as greatly to shorten distances?[2]

The discovery of the steam-engine owed its birth, like most human inventions, to rude attempts which have been attributed to different persons, while the real author is not certainly known. It is, however, less in the first attempts that the principal discovery consists, than in the successive improvements which have brought steam-engines to the conditions in which we find them today. There is almost as great a distance between the first apparatus in which the expansive force of steam was displayed and the existing machine, as between the first raft that man ever made and the modern vessel.

If the honor of a discovery belongs to the nation in which it has acquired its growth and all its developments, this honor cannot be here refused to England. Savery, Newcomen, Smeaton, the famous Watt, Woolf, Trevithick, and some other English engineers, are the veritable creators of the steam-engine. It has acquired at their hands all its successive degrees of improvement. Finally, it is natural that an invention should have its birth and especially be developed, be perfected, in that place where its want is most strongly felt.

Notwithstanding the work of all kinds done by steam-engines, notwithstanding the satisfactory condition to which they have been brought today, their theory is very little understood, and the attempts to improve them are still directed almost by chance.

The question has often been raised whether the motive power of heat[3] is unbounded, whether the possible improvements in steam-engines have an assignable limit—a limit which the nature of things will not allow to be passed by any means whatever; or whether, on the contrary, these improvements may be carried on indefinitely. We have long sought, and are seeking today, to ascertain whether there are in existence agents preferable to the vapor of water for developing the motive power of heat; whether atmospheric air, for example, would not present in this respect great advantages. We propose now to submit these questions to a deliberate examination.

The phenomenon of the production of motion by heat has not been considered from a sufficiently general point of view. We have considered it only in machines the nature and mode of action of which have not allowed us to take in the whole extent of application of which it is susceptible. In such machines the phenomenon is, in a way, incomplete. It becomes difficult to recognize its principles and study its laws.

In order to consider in the most general way the principle of the production of motion by heat, it must be considered independently of any mechanism or any particular agent. It is necessary to establish principles applicable not only to steam-engines[4] but to all imaginable heat-engines, whatever the working substance and whatever the method by which it is operated.

Machines which do not receive their motion from heat, those which have for a motor the force of men or of animals, a waterfall, an air current, etc., can be studied even to their smallest details by the mechanical theory. All cases are foreseen, all imaginable movements are referred to these general principles, firmly established, and applicable under all circumstances. This is the character of a complete theory. A similar theory is evidently needed for heat-engines. We shall have it only when the laws of physics shall be extended enough, generalized enough, to make known beforehand all the effects of heat acting in a determined manner on any body.

We will suppose in what follows at least a superficial knowledge of the different parts which compose an ordinary steam-engine; and we consider it unnecessary to explain what are the furnace, boiler, steam-cylinder, piston, condenser, etc.

The production of motion in steam-engines is always accompanied by a circumstance on which we should fix our attention. This circumstance is the re-establishing of equilibrium in the caloric; that is, its passage from a body in which the temperature is more or less elevated, to another in which it is lower. What happens in fact in a steam-engine actually in motion? The caloric developed in the furnace by the effect of the combustion traverses the walls of the boiler, produces steam, and in some way incorporates itself with it. The latter carrying it away, takes it first into the cylinder, where it performs some function, and from thence into the condenser, where it is liquefied by contact with the cold water which it encounters there. Then, as a final result, the cold water of the condenser takes possession of the caloric developed by the combustion. It is heated by the intervention of the steam as if it had been placed directly over the furnace. The steam is here only a means of transporting the caloric. It fills the same office as in the heating of baths by steam, except that in this case its motion is rendered useful.

We easily recognize in the operations that we have just described the re-establishment of equilibrium in the caloric, its passage from a more or less heated body to a cooler one. The first of these bodies, in this case, is the heated air of the furnace; the second is the condensing water. The re-establishment of equilibrium of the caloric takes place between them, if not completely, at least partially, for on the one hand the heated air, after having performed its function, having passed round the boiler, goes out through the chimney with a temperature much below that which it had acquired as the effect of combustion; and on the other hand, the water of the condenser, after having liquefied the steam, leaves the machine with a temperature higher than that with which it entered.

The production of motive power is then due in steam-engines not to an actual consumption of caloric, but *to its transportation from a warm body to a cold body*, that is, to its re-establishment of equilibrium—an equilibrium considered as destroyed by any cause whatever, by chemical action such as combustion, or by any other. We shall see shortly that this principle is applicable to any machine set in motion by heat.

According to this principle, the production of heat alone is not sufficient to give birth to the impelling power: it is necessary that there should also be cold; without it, the heat would be useless. And in fact, if we should find about us only bodies as hot as our furnaces, how can we condense steam? What should we do with it if once produced? We should not presume that we might discharge it into the atmosphere, as is done in some engines;[5] the atmosphere would not receive it. It does receive it under the actual condition of things, only because it fulfils the office of a vast condenser, because it is at a lower temperature; otherwise it would soon become fully charged, or rather would be already saturated.[6]

Wherever there exists a difference of temperature, wherever it has been possible for the equilibrium of the caloric to be re-established, it is possible to have also the production of impelling power. Steam is a means of realizing this power, but it is not the only one. All substances in nature can be employed for this purpose, all are susceptible of changes of volume, of successive contradictions and dilatations, through the alternation of heat and cold. All are capable of overcoming in their changes of volume certain resistances, and of thus developing the impelling power. A solid body—a metallic bar for example—alternately heated and cooled increases and diminishes in length, and can move bodies fastened to its ends. A liquid alternately heated and cooled increases and diminishes in volume, and can overcome obstacles of greater or less size, opposed to its dilatation. An aeriform fluid is susceptible of considerable change of volume by variations of temperature. If it is enclosed in an expansible space, such as a cylinder provided with a piston, it will produce movements of great extent. Vapors of all substances capable of passing into a gaseous condition, as of alcohol, of mercury, of sulphur, etc., may fulfil the same office as vapor of water. The latter, alternately heated and cooled, would produce motive power in the shape of permanent gases, that is, without ever returning to a liquid state. Most of these substances have been proposed, many even have been tried, although up to this time perhaps without remarkable success.

We have shown that in steam-engines the motive-power is due to a re-establishment of equilibrium in the caloric; this takes place not only for steam-engines, but also for every heat-engine—that is, for every machine of which caloric is the motor. Heat can evidently be a cause of motion only by virtue of the changes of volume or of form which it produces in bodies.

These changes are not caused by uniform temperature, but rather by alternations of heat and cold. Now to heat any substance whatever requires a body warmer than the one to be heated; to cool it requires a cooler body. We supply caloric to the first of these bodies that we may transmit it to the second by means of the intermediary substance. This is to re-establish, or at least to endeavor to re-establish, the equilibrium of the caloric.

It is natural to ask here this curious and important question: Is the motive power of heat invariable in quantity, or does it vary with the agent employed to realize it as the intermediary substance, selected as the subject of action of the heat?

It is clear that this question can be asked only in regard to a given quantity of caloric,[7] the difference of the temperatures also being given. We take, for example, one body A kept at a temperature of 100° and another body B kept at a temperature of 0°, and ask what quantity of motive power can be produced by the passage of a given portion of caloric (for example, as much as is necessary to melt a kilogram of ice; from the first of these bodies to the second. We inquire whether this quantity of motive power is necessarily limited, whether it varies with the substance employed to realize whether the vapor of water offers in this respect more or less advantage than the vapor of alcohol, of mercury, a permanent gas, or any other substance. We will try to answer these questions, availing ourselves of ideas already established.

We have already remarked upon this self-evident fact, or fact which at least appears evident as soon as we reflect on the changes volume occasioned by heat: *wherever there exists a difference of temperature motive power can be produced.* Reciprocally, wherever we can resume this power, it is possible to produce a difference of temperature, it is possible to occasion destruction of equilibrium in the caloric. Are not percussion and the friction of bodies actually means of raising their temperature, of making it reach spontaneously a higher degree than that of the surrounding bodies, and consequently of producing a destruction of equilibrium in the caloric, where equilibrium previously existed? It is a fact proved by experience, that the temperature of gaseous fluids is raised by compression and lowered by rarefaction. This is a sure method of changing the temperature of bodies, and destroying the equilibrium of the caloric as many times as may be desired with the same substance. The vapor of water employed in an inverse manner to that in which it is used in steam-engines can also be regarded as a means of destroying the equilibrium of the caloric. To be convinced of this we need to observe closely the manner in which motive power is developed by the action of heat on vapor of water. Imagine two bodies *A* and *B*, kept each at a constant temperature, that of *A* being higher than that of *B*. These two bodies, to which we can give or from which we can remove the heat without causing their temperatures to vary, exercise the functions of two unlimited reservoirs of caloric. We will call the first the furnace and the second the refrigerator.

If we wish to produce motive power by carrying a certain quantity of heat from the body *A* to the body *B* we shall proceed as follows:[8]

(1) To borrow caloric from the body *A* to make steam with it—that is, to make this body fulfil the function of a furnace, or rather of the metal composing the boiler in ordinary engines—we here assume that the steam is produced at the same temperature as the body *A*.

(2) The steam having been received in a space capable of expansion, such as a cylinder furnished with a piston, to increase the volume of this space, and consequently also that of the steam. Thus rarefied, the temperature will fall spontaneously, as occurs with all elastic fluids; admit that the rarefaction may be continued to the point where the temperature becomes precisely that of the body *B*.

(3) To condense the steam by putting it in contact with the body *B*, and at the same time exerting on it a constant pressure until it is entirely liquefied. The body *B* fills here the place of the injection water in ordinary engines, with this difference, that it condenses the vapor without mingling with it, and without changing its own temperature.[9]

Notes

1. It may be said that coal-mining has increased tenfold in England since the invention of the steam-engine. It is almost equally true in regard to the mining of copper, tin, and iron. The results produced in a half-century by the steam-engine in the mines of England are today paralleled in the gold and silver mines of the New World—mines of which the working declined from day to day, principally on account of the insufficiency of the motors employed in the draining and the extraction of the minerals.

2. We say, to lessen the dangers of journeys. In fact, although the use of the steam-engine on ships is attended by some danger which has been greatly exaggerated, this is more than compensated by the power of following always an appointed and well-known route, of resisting the force of the winds which would drive the ship towards the shore, the shoals, or the rocks.

3. We use here the expression motive power to express the useful effect that a motor is capable of producing. This effect can always be likened to the elevation of a weight to a certain height. It has, as we know, as a measure, the product of the weight multiplied by the height to which it is raised.

4. We distinguish here the steam-engine from the heat-engine in general. The latter may make use of any agent whatever, of the vapor of water or of any other, to develop the motive power of heat.

5. Certain engines at high pressure throw the steam out into the atmosphere instead of the Condenser. They are used specially in places where it would be difficult to procure a stream of cold water sufficient to produce condensation.

6. The existence of water in the liquid state here necessarily assumed, since without it the steam-engine could not be fed, supposes the existence of a pressure capable of preventing this water from vaporizing, consequently of a pressure equal or superior to the tension of vapor at that temperature. If such a pressure were not exerted by the atmospheric air, there would be instantly produced a quantity of steam sufficient to give rise to that tension, and it would be necessary always to overcome this pressure in order to throw out the steam from the engines into the new atmosphere. Now this is evidently equivalent to overcoming the tension which the steam retains after its condensation, as effected by ordinary means.

7. If a very high temperature existed at the surface of our globe, as it seems certain that it exists in its interior, all the waters of the ocean would be in a state of vapor in the atmosphere, and no portion of it would be found in a liquid state.

8. It is considered unnecessary to explain here what is quantity of caloric or quantity of heat (for we employ these two expressions indifferently), or to describe how we measure these quantities by the calorimeter. Nor will we explain what meant by latent heat, degree of temperature, specific heat, etc. The reader could be familiarized with these terms through the study of the elementary arises of physics or of chemistry.

9. [This is only a sketch and Carnot accidentally leaves the cycle incomplete. E. M.]

10. We may perhaps wonder here that the body B being at the same temperature as the steam is able to condense it. Doubtless this is not strictly possible, but the slightest difference of temperature will determine the condensation, which suffices.

CHAPTER 5
ELECTRICITY AND MAGNETISM

*Electricity and magnetism are two aspects of one force—
the electromagnetic force.*

INTRODUCTION

BENJAMIN FRANKLIN (1706–1790), born in Boston, was the first American to make major contributions to science. After a successful career as a printer in Philadelphia, Franklin found time to engage in an amazing number of activities, from his role in the founding of America to his pioneering work in the study of electricity. The following two short letters document two of his most important discoveries. The first is the letter that established the fact that although electricity comes in positive and negative forms, only one of those forms actually moves. (Today, we would say that it is the movement of electrons that usually accounts for electrical charge). The second letter, coupled with a later descriptive account by British chemist Joseph Priestley (1733–1804), recounts his famous kite experiment, which established the electrical nature of lightning. (Incidentally, this experiment is potentially dangerous and should NOT be repeated by individuals).

 Experiments and Observation on Electricity, Made at Philadelphia in America

by Benjamin Franklin (1751)

LETTER ON THE ONE-FLUID THEORY OF ELECTRICITY

1. A PERSON standing on wax, and rubbing a tube, and another person on wax drawing the fire; they will both of them, provided they do not stand so as to touch one another, appear to be electrified to a person standing on the floor; that is, he will perceive a spark on approaching each of them with his knuckle.

2. But if the persons on wax touch one another during the exciting of the tube, neither of them will appear to be electrified.

3. If they touch one another after the exciting of the tube and drawing the fire as aforesaid, there will be a stronger spark between them than there was between either of them and the person on the floor.

4. After such a strong spark neither of them discover any electricity.

These appearances we attempt to account for thus:

We suppose, as aforesaid, that electrical fire is a common element, of which every one of these three persons has his equal share before any operation is begun with the tube. A, who stands upon wax, and rubs the tube, collects the electrical fire from himself into the glass; and his communication with the common stock being cut off by the wax, his body is not again immediately supplied. B, who stands upon wax likewise, passing his knuckle along the tube, receives the fire which was collected from the glass by A; and his communication with the common stock being cut off, he retains the additional quantity received. To C standing on the floor, both appear to be electrified; for he, having only the middle quantity of electrical fire, receives a spark upon approaching B, who has an over quantity, but gives one to A, who has an under quantity. If A and B approach to touch each other, the spark is stronger; because the difference between them is greater. After such touch, there is no spark between either of them and C, because the electrical fire in all is reduced to the original equality. If they touch while electrising, the equality is never destroyed, the fire only circulating. Hence have arisen some new terms among us. We say, B (and bodies alike circumstanced) is electrised positively; A, negatively; or rather B is electrised plus, A, minus. And we daily in our experiments electrise plus or minus, as we think proper. To electrise plus or minus, no more needs be known than this; that the parts of the tube or sphere that are rubbed, do in the instant of the friction attract the electrical fire, and therefore take it from the thing rubbing. The same parts immediately, as the friction upon them ceases, are disposed to give the fire, they have received, to any body that has less.

A Letter of Benjamin Franklin, Esq; to Peter Collinson F. R. S. Concerning an Electrical Kite

by Benjamin Franklin (1752)

SIR:—AS FREQUENT MENTION is made in publick papers from Europe of the success of the Philadelphia experiment for drawing the electric fire from clouds by means of pointed rods of iron erected on high buildings, &c., it may be agreeable to the curious to be informed that the same experiment has succeeded in Philadelphia, though made in a different and more easy manner, which is as follows.

Make a small cross of two light strips of cedar, the arms so long as to reach to the four corners of a large thin silk handkerchief when extended; tie the corners of the handkerchief to the extremities of the cross, so you have the body of a kite; which, being properly accommodated with a tail, loop, and string, will rise in the air, like those made of paper; but this being silk is fitter to bear the wet and wind of a thunder-gust without tearing. To the top of the upright stick of the cross is to be fixed a very sharp-pointed wire, rising a foot or more above the wood. To the end of the twine, next the hand, is to be tied a silk ribbon, and where the silk and twine join, a key may be fastened. This kite is to be raised when a thunder-gust appears to be coming on, and the person who holds the string must stand within a door or window, or under some cover, so that the silk ribbon may not be wet; and care must be taken that the twine does not touch the frame of the door or window. As soon as any of the thunder-clouds come over the kite, the pointed wire will draw the electric fire from them, and the kite, with all the twine, will be electrified, and the loose filaments of the twine will stand out every way, and be attracted by an approaching finger. And when the rain has wetted the kite and twine, so that it can conduct the electric fire freely, you will find it stream out plentifully from the key on the approach of your knuckle. At this key the phial may be charged; and from electric fire thus obtained spirits may be kindled, and all the other electric experiments be performed which are usually done by the help of a rubbed glass globe or tube, and thereby the sameness of the electric matter with that of lightning completely demonstrated.

B. Franklin
Philadelphia, 19 October, 1752

The History and Present State of Electricity, with Original Experiments

by Joseph Priestley (1775)

As every circumstance relating to so capital a discovery as this (the greatest, perhaps, that has been made in the whole compass of philosophy, since the time of Sir Isaac Newton) cannot but give pleasure to all my readers, I shall endeavour to gratify them with the communication of a few particulars which I have from the best authority.

The Doctor, after having published his method of verifying his hypothesis concerning the sameness of electricity with the matter of lightning, was waiting for the erection of a spire in Philadelphia to carry his views into execution; not imagining that a pointed rod, of a moderate height, could answer the purpose; when it occurred to him, that, by means of a common kite, he could have a readier and better access to the regions of thunder than by any spire whatever. Preparing, therefore, a large silk handkerchief, and two cross sticks, of a proper length, on which to extend it, he took the opportunity of the first approaching thunder storm to take a walk into a field, in which there was a shed convenient for his purpose. But dreading the ridicule which too commonly attends unsuccessful attempts in science, he communicated his intended experiment to no body but his son, who assisted him in raising the kite.

The kite being raised, a considerable time elapsed before there was any appearance of its being electrified. One very promising cloud had passed over it without any effect; when, at length, just as he was beginning to despair of his contrivance, he observed some loose threads of the hempen string to stand erect, and to avoid one another, just as if they had been suspended on a common conductor. Struck with this promising appearance, he immediately presented his knuckle to the key, and (let the reader judge of the exquisite pleasure he must have felt at that moment) the discovery was complete. He perceived a very evident electric spark. Others succeeded, even before the string was wet, so as to put the matter past all dispute, and when the rain had wetted the string, he collected electric fire very copiously. This happened in June 1752, a month after the electricians in France had verified the same theory, but before he had heard of any thing that they had done.

Besides this kite, Dr. Franklin had afterwards an insulated iron rod to draw the lightning into his house, in order to make experiments whenever there should be a considerable quantity of it in the atmosphere; and that he might not lose any opportunity of that nature, he connected two bells with this apparatus, which gave him notice, by their ringing, whenever his rod was electrified.

On the Electricity Excited by the Mere Contact of Conducting Substances of Different Kinds

by Alessandro Volta (1800)

ALESSANDRO VOLTA (1745–1627) was born in Como, Italy, and spent most of his academic career at the University of Padua. His most important contribution to science was the development of what he called the voltaic pile and we call a battery. Before this invention, the only way to study electricity was to build up charge on an object by rubbing it. A common device, called a Leiden jar, stored electrical charge in large glass jars. With

Volta's discovery, it became possible to produce electricity easily and, more important, to study the effects of moving electrical charges (current). This led to the discovery of the fundamental connection between electricity and magnetism.

* * *

IN A LETTER FROM MR. ALEXANDER VOLTA, F. R. S. PROFESSOR OF NATURAL PHILOSOPHY IN THE UNIVERSITY OF PAVIA, TO THE RT. HON. SIR JOSEPH BANKS, BART. K. B. P. R. S.

After a long silence, which I do not attempt to excuse, I have the pleasure of communicating to you, Sir, and through you to the Royal Society, some striking results to which I have come in carrying out my experiments on electricity excited by the simple mutual contact of metals of different sorts, and even by the contact of other conductors, also different among themselves, whether liquids or containing some liquid, to which property they owe their conducting power. The most important of these results, which includes practically all the others, is the construction of an apparatus which, in the effects which it produces, that is, in the disturbances which it produces in the arms etc., resembles Leyden jars, or better still electric batteries feebly charged, which act unceasingly or so that their charge after each discharge reestablishes itself; which in a word provides an unlimited charge or imposes a perpetual action or impulsion on the electric fluid; but which otherwise is essentially different from these, both because of this continued action which is its property and because, instead of being made, as are the ordinary jars and electric batteries, of one or more insulating plates in thin layers of those bodies which are thought to be the only electric bodies, coated with conductors or bodies called non-electrics, this new apparatus is formed altogether of several of these latter bodies, chosen even among the best conductors and therefore the most remote, according to what has always been believed, from the electric nature. Yes, the apparatus of which I speak, and which will doubtless astonish you, is only an assemblage of a number of good conductors of different sorts arranged in a certain way. 30, 40, 60, pieces or more of copper, or better of silver, each in contact with a piece of tin, or what is much better, of zinc and an equal number of layers of water or some other liquid which is a better conductor than pure water, such as salt-water or lye and so forth, or pieces of cardboard or of leather, etc. well soaked with these liquids; when such layers are interposed between each couple or combination of the two different metals, such an alternative series of these three sorts of conductors always in the same order, constitutes my new instrument; which imitates, as I have said, the effects of Leyden jars or of electric batteries by giving the same disturbances as they; which in truth, are much inferior to these batteries when highly charged in the force and noise of their explosions, in the spark, in the distance through which the charge can pass, etc., and equal in effect only to a battery very feebly charged, but a battery nevertheless of an immense capacity;

but which further infinitely surpasses the power of these batteries in that it does not need, as they do, to be charged in advance by means of an outs de source; and in that it can give the disturbance every time that it is properly touched, no matter how often.

I proceed to give a more detailed description of this apparatus and of some other analogous ones, as well as the most remarkable experiments made with them.

I provided myself with several dozen small round plates or discs of copper, of brass, or better of silver, an inch in diameter more or less (for example, coins) and an equal number of plates of tin, or which is much better, of zinc, approximately of the same shape and size; I say approximately because precision is not necessary, and in general the size as well as the shape of the metallic pieces is arbitrary: all that is necessary is that they may be arranged easily one above the other in a column. I further provided a sufficiently large number of discs of cardboard, of leather, or of some other spongy matter which can take up and retain much water, or the liquid with which they must be well made if the experiment is to succeed. These pieces, which I will call the moistened discs, I make a little smaller than the metallic discs or plates, so that when placed between them in the way that I shall soon describe, they do not protrude.

Now having in hand all these pieces in good condition, that is to say, the metallic discs clean and dry, and the other non-metallic ones well soaked in water or which is much better, in brine, and afterwards slightly wiped so that the liquid does not come out in drops, I have only to arrange them in the proper way; and this arrangement is simple and easy.

I place horizontally on a table or base one of the metallic plates, for example, one of the silver ones, and on this first plate I place a second plate of zinc; on this second plate I lay one of the moistened discs; then another plate of silver, followed immediately by another of zinc, on which I place again a moistened disc. I thus continue in the same way coupling a plate of silver with one of zinc, always in the same sense, that is to say, always silver below and zinc, above or *vice versa*, according as I began, and inserting between these couples a moistened disc; I continue, I say, to form from several of these steps a column as high as can hold itself up without falling. … [*Volta goes on to describe other versions of this primitive battery*]

Experimental Researches in Electricity
by Michael Faraday (1832)

MICHAEL FARADAY (1792–1867) was born in London. His family was not well off, and, since they were members of a small Protestant sect and not members of the Church of England, the normal educational system was not available to him. He was apprenticed to a bookbinder, and educated himself by reading the books he was binding. He attended a series of lectures by Humphrey Davy, one of England's outstanding chemists, and sent Davy his lecture notes, beautifully bound, as a calling card. Soon after that, he was hired by Davy, first as a secretary, then as an assistant. He quickly rose in the scientific world, becoming the outstanding physicist of his generation, a man widely acclaimed around the world. He was a frequent visitor in the court of Queen Victoria and his public lectures attracted thousands of listeners. Faraday was a hands-on kind of scientist, always aware of his lack of background in theory. You can get some sense of that from the following excerpt, in which he describes some experiments in electromagnetic induction. Remember as you read this that it was this sort of experiment that led to the commercial generation of electricity, which is the basis of our modern technological society.

* * *

2. EVOLUTION OF ELECTRICITY FROM MAGNETISM

27. A welded ring was made of soft round bar-iron, the metal being seven eighths of an inch in thickness, and the ring six inches in external diameter. Three helices were put round one part of this ring, each containing about twenty-four feet of copper wire one twentieth of an inch thick; they were insulated from the iron and each other, and superposed in the manner before described (6), occupying about nine inches in length upon the ring. They could be used separately or conjointly; the group may be distinguished by the letter A (Fig. 91, 1). On the other part of the ring about sixty feet of similar copper wire in two pieces were applied in the same manner, forming a helix B, which had the same common direction with the helices of A, but being separated from it at each extremity by about half an inch of the uncovered iron.

$$X = Ne\left(\frac{1}{r^2} - \frac{r}{R^3}\right)$$

$$V = Ne\left(\frac{1}{r} - \frac{3}{2R} + \frac{r^2}{2R^3}\right).$$

28. The helix *B* was connected by copper wires with a galvanometer three feet from the ring. The helices of A were connected end to end so as to form one common helix, the extremities of which were connected with a battery of ten pairs of plates four inches square. The galvanometer was immediately affected, and to a degree far beyond what has been described

when with a battery of tenfold power helices *without iron* were used (10); but though the contact was continued, the effect was not permanent, for the needle soon came to rest in its natural position, as if quite indifferent to the attached electro-magnetic arrangement. Upon breaking the contact with the battery, the needle was again powerfully deflected, but in the contrary direction to that induced in the first instance.

29. Upon arranging the apparatus so that B should be out of use, the galvanometer be connected with one of the three wires of A (27), and the other two made into a helix through which the current from the trough (28) was passed, similar but rather more powerful effects were produced.

30. When the battery contact was made in one direction, the galvanometer-needle was deflected on the one side; if made in the other direction, the deflection was on the other side. The deflection on breaking the battery contact was always the reverse of that produced by completing it. The deflection on making a battery contact always indicated an induced current in the opposite direction to that from the battery; but on breaking the contact the deflection indicated an induced current in the same direction as that of the battery. No making or breaking of the contact at B side, or in any part of the galvanometer circuit, produced any effect at the galvanometer. No continuance of the battery current caused any deflection of the galvanometer-needle. As the above results are common to all these experiments, and to similar ones with ordinary magnets to be hereafter detailed, they need not be again particularly described.

31. Upon using the power of one hundred pairs of plates (10), with this ring, the impulse at the galvanometer, when contact was completed or broken, was so great as to make the needle spin round rapidly four or five times, before the air and terrestrial magnetism could reduce its motion to mere oscillation. ...

34. Another arrangement was then employed connecting the former experiments on volta-electric induction (6–26) with the present. A combination of helices like that already described (6) was constructed upon a hollow cylinder of pasteboard: there were eight lengths of copper wire, containing altogether 220 feet; four of these helices were connected end to end, and then with the galvanometer (7); the other intervening four were also connected end to end, and the battery of one hundred pairs discharged through them. In this form the effect on the galvanometer was hardly sensible (11), though magnets could be made by the induced current (13). But when a soft iron cylinder seven eighths of an inch thick, and twelve inches long, was introduced into the pasteboard tube, surrounded by the helices, then the induced current affected the galvanometer powerfully, and with all the phenomena just described (30). It possessed also the power of making magnets with more energy, apparently, than when no iron cylinder was present. ...

The experiments described combine to prove that when a piece of metal (and the same may be true of all conducting matter (213)) is passed either before a single pole, or between the opposite poles of a magnet, or near electro-magnetic poles, whether ferruginous or not, electrical currents are produced across the metal transverse to the direction of motion; and which therefore, in Arago's experiments, will approximate towards the direction of Radii. If a single wire be moved like the spoke of a wheel near a magnetic pole, a current of electricity is determined through it from one end towards the other. If a wheel be imagined, constructed of a great number of these radii, and this revolved near the pole, in the manner of the copper disc (85), each radius will have a current produced in it as it passes by the pole. If the radii be supposed to be in contact laterally, a copper disc results, in which the directions of the currents will be generally the same, being modified only by the coaction which can take place between the particles, now that they are in metallic contact.

CHAPTER 6
WAVES AND
ELECTROMAGNETIC RADIATION

Whenever an electrically charged object is accelerated, it produces electromagnetic radiation—waves of energy that travel at the speed of light.

INTRODUCTION

H EINRICH HERTZ (1857–1894) was born to a prominent family in Hamburg and, as was the custom at the time, studied at several German universities before taking his doctorate in theoretical physics at Berlin. He eventually became a professor at the University of Karlsruhe. In 1887 he published experiments that established the existence of electromagnetic waves—what we would call radio waves today. In this excerpt, he examines the properties of the newly discovered radiation. One interesting aspect of this work is that Hertz had absolutely no idea of the enormous practical implications of his work—in fact, he said that "It is of no use whatever." A few years later, of course, Guilgelmo Marconi perfected what we would now call a radio communication system. Hertz died of an infection at the age of 36.

Electric Radiation
by Heinrich Hertz (1887)

AS SOON AS I had succeeded in proving that the action of an electric oscillation spreads out as a wave into space, I planned experiments with the object of concentrating this action and making it perceptible at greater distances by putting the primary conductor in the focal line of a large concave parabolic mirror. These experiments did not lead to the desired result, and I felt certain that the want of success was a necessary consequence of the disproportion between the length (4–5 metres) of the waves used and the dimensions which I was able, under the most favourable circumstances, to give to the mirror. Recently I have observed that the experiments which I have described can be carried out quite well with oscillations of more than ten times the frequency, and with waves less than one-tenth the length of those which were first discovered. I have, therefore, returned to the use of concave mirrors, and have obtained better results than I had ventured to hope for. I have succeeded in producing distinct rays of electric force, and in carrying out with them the elementary experiments which are commonly performed with light and radiant heat. The following is an account of these experiments:

THE APPARATUS

The short waves were excited by the same method which we used for producing the longer waves. The primary conductor used may be most simply described as follows: Imagine a cylindrical brass body, 3 cm. in diameter and 26 cm. long, interrupted midway along its length by a spark-gap whose poles on either side are formed by spheres of 2 cm. radius. The length of the conductor is approximately equal to the half wave-length of the corresponding oscillation in straight wires; from this we are at once able to estimate approximately the period of oscillation. It is essential that the pole-surfaces of the spark-gap should be frequently repolished, and also that during the experiments they should be carefully protected from illumination by simultaneous side-discharges; otherwise the oscillations are not excited. Whether the spark-gap is in a satisfactory state can always, be recognized by the appearance and sound of the sparks. The discharge is led to the two halves of the conductor by means of two gutta-percha-covered wires which are connected near the spark-gap on either side.

Here, again, the small sparks induced in a secondary conductor were the means used for detecting the electric forces in space. As before, I used partly a circle which could be rotated within itself and which had about the same period of oscillation as the primary conductor. It was made of copper wire 1 mm. thick, and had in the present instance a diameter of only 7.5 cm. One end of the wire carried a polished brass sphere a few millimetres in diameter; the other end was pointed and could be brought up, by means of a fine screw insulated from the wire, to within an exceedingly short distance from the brass sphere. As will be readily understood, we have here to deal only with minute sparks of a few hundredths of a millimetre in length; and after a little practice one judges more according to the brilliancy than the length of the sparks.

THE PRODUCTION OF THE RAY

If the primary oscillator is now set up in a fairly large free space, one can, with the aid of the circular conductor, detect in its neighbourhood on a smaller scale all those phenomena which I have already observed and described as occurring in the neighbourhood of a larger oscillation. The greatest distance at which sparks could be perceived in the secondary conductor was 1.5 metre, or, when the primary spark-gap was in very good order, as much as 2 metres.

RECTILINEAR PROPAGATION

If a screen of sheet zinc 2 metres high and 1 metre broad is placed on the straight line joining both mirrors, and at right angles to the direction of the ray, the secondary sparks disappear completely. An equally complete shadow is thrown by a screen of tinfoil or gold-paper. If an assistant walks across the path of the ray, the secondary spark-gap becomes dark as soon as he intercepts the ray, and again lights up when he leaves the path clear. Insulators do not stop the ray—it passes right through a wooden partition or door; and it is not without astonishment that one sees the sparks appear inside a closed room.

[*Hertz then describes several experiments on the polarization, reflection, and refraction of the 'rays'—the following gives the flavor of the work.*]

REFRACTION

In order to find out whether any refraction of the ray takes place in passing from air into another insulating medium, I had a large prism made of so-called hard pitch, a material like asphalt. The base was an isosceles triangle 1.2 metres in the side, and with a refracting angle of nearly 30°. The refracting edge was placed vertical, and the height of the whole prism was 1.5 metres. But since the prism weighed about 12 cwt., and would have been too heavy to move as a whole, it was built up of three pieces, each 0.5 metre high, placed one above the other. The material was cast in wooden boxes which were left around it, as they did not appear to interfere with its use. The prism was mounted on a support of such height that the middle of its refracting edge was at the same height as the primary and secondary spark-gaps. When I was satisfied that refraction did take place, and had obtained some idea of its amount, I arranged the experiment in the following manner: The producing mirror was set up at a distance of 2.6 metres from the prism and facing one of the refracting surfaces, so that the axis of the beam was directed as nearly as possible towards the centre of mass of the prism, and met the refracting surface at an angle of incidence of 25° (on the side of the normal towards the base). Near the refracting edge and also at the opposite side of the prism were placed two conducting screens which prevented the ray from passing by any other path than that through the prism. On the side of the emerging ray there was marked upon the floor a circle of 2.5 metres radius, having as its centre the centre of mass of the lower end of the prism. Along this the receiving mirror was now moved about, its aperture being always directed towards the centre of the circle. No sparks were obtained when the mirror was placed in the direction of the incident ray produced; in this direction the prism threw a complete shadow. But sparks appeared when the mirror was moved towards the base of the prism, beginning when the angular deviation from the first position was about 11°. The sparking increased in intensity until the deviation amounted to about 22°, and then again decreased. The last sparks were observed with a deviation of about 34°. When the mirror was placed in a position of maximum effect, and then moved away from the prism along the radius of the circle, the sparks could be traced up to a distance of 5–6 metres. When an assistant stood either in front of the prism or behind it the sparking invariably ceased, which shows that the action reaches the secondary conductor through the prism and not in any other way. The experiments were repeated after placing both mirrors with their focal lines horizontal, but without altering the position of the prism. This made no difference in the phenomena observed. A refracting angle of 30° and a deviation of 22° in the neighbourhood of the minimum deviation corresponds to a refractive index of 1.69. The refractive index of pitch-like materials for light is given as being between 1.5 and 1.6. We must not attribute any importance to the magnitude or even the sense of this difference, seeing that our method was not an accurate one, and that the material used was impure.

We have applied the term rays of electric force to the phenomena which we have investigated. We may perhaps further designate them as rays of light of very great wave-length. The experiments described appear to me, at any rate, eminently adapted to remove any doubt as to the identity of light, radiant heat, and electromagnetic wave-motion. I believe that from now on we shall have greater confidence in making use of the advantages which this identity enables, us to derive both in the study of optics and of electricity.

CHAPTER 7
THE THEORY OF RELATIVITY

All observers, no matter what their frame of reference,
see the same laws of nature.

INTRODUCTION

ALBERT EINSTEIN (1879–1955) was born in Ulm, in what is now Germany. There is a lot of folklore about his student days, so for the record let us state that (1) he did *not* flunk mathematics, and (2) he did *not* flunk out of high school (although he did have issues with insubordination in the strict German system). He took his doctorate at Zurich and, unable to find the teaching job he was looking for, accepted a job with the Swiss Patent Office at Bern—probably the most unusual appointment in the history of science. There, in 1905, he published three extraordinary papers, each of which could have established his reputation and which, taken together, explain why physicists refer to 1905 as the 'Miraculous Year.' One paper, on Brownian motion, established the physical existence of atoms and another, on the photoelectric effect, was crucial to the development of quantum mechanics. (It was, in fact, this latter paper that was the basis of his Nobel Prize in 1921). But of course, his most famous work is presented in the third paper, on the special theory of relativity. In the excerpt below, Einstein explains this theory in non-mathematical terms.

He quickly entered the world of German academic physics, rising to the rank of professor in Berlin. With the threat of the Nazis on the horizon, he came to America in 1932, eventually becoming one of the first members of the Institute for Advanced Study at Princeton, a position he occupied until his death. His status as both a scientist and an icon of popular culture was exemplified by his nomination as "Man of the Century" by *Time* magazine in 1999.

Relativity: The Special and General Theory
by Albert Einstein (1920)

I. PHYSICAL MEANING OF GEOMETRICAL PROPOSITIONS

In your schooldays most of you who read this book made acquaintance with the noble building of Euclid's geometry, and you remember—perhaps with more respect than love—the magnificent structure, on the lofty staircase of which you were chased about for uncounted hours by conscientious teachers. By reason of your past experience, you would certainly regard every one with disdain who should pronounce even the most out-of-the-way proposition of this science to be untrue. But perhaps this feeling of proud certainty would leave you immediately if some one were to ask you: "What, then, do you mean by the assertion that these propositions are true?" Let us proceed to give this question a little consideration.

Geometry sets out from certain conceptions such as "plane," "point," and "straight line," with which we are able to associate more or less definite ideas, and from certain simple propositions (axioms) which, in virtue of these ideas, we are inclined to accept as "true." Then, on the basis of a logical process, the justification of which we feel ourselves compelled to admit, all remaining propositions are shown to follow from those axioms, *i.e.* they are proven. A proposition is then correct ("true") when it has been derived in the recognised manner from the axioms. The question of the "truth" of the individual geometrical propositions is thus reduced to one of the "truth" of the axioms. Now it has long been known that the last question is not only unanswerable by the methods of geometry, but that it is in itself entirely without meaning. We cannot ask whether it is true that only one straight line goes through two points. We can only say that Euclidean geometry deals with things called "straight line," to each of which is ascribed the property of being uniquely determined by two points situated on it. The concept "true" does not tally with the assertions of pure geometry, because by the word "true" we are eventually in the habit of designating always the correspondence with a "real" object; geometry, however, is not concerned with the relation of the ideas involved in it to objects of experience, but only with the logical connection of these ideas among themselves.

It is not difficult to understand why, in spite of this, we feel constrained to call the propositions of geometry "true." Geometrical ideas correspond to more or less exact objects in nature, and these last are undoubtedly the exclusive cause of the genesis of those ideas. Geometry ought to refrain from such a course, in order to give to its structure the largest possible logical unity. The practice, for example, of seeing in a "distance" two marked positions on a practically rigid body is something which is lodged deeply in our habit of thought. We are accustomed further to regard three points as being situated on a straight line, if their apparent positions can be made to coincide for observation with one eye, under suitable choice of our place of observation.

If, in pursuance of our habit of thought, we now supplement the propositions of Euclidean geometry by the single proposition that two points on a practically rigid body always correspond to the same distance (line-interval), independently of any changes in position to which we may subject the body, the propositions of Euclidean geometry then resolve themselves into propositions on the possible relative position of practically rigid bodies. Geometry which has been supplemented in this way is then to be treated as a branch of physics. We can now legitimately ask as to the "truth" of geometrical propositions interpreted in this way, since we are justified in asking whether these propositions are satisfied for those real things we have associated with the geometrical ideas. In less exact terms we can express this by saying that by the "truth" of a geometrical proposition in this sense we understand its validity for a construction with ruler and compasses.

Of course the conviction of the "truth" of geometrical propositions in this sense is founded exclusively on rather incomplete experience. For the present we shall assume the "truth" of the geometrical propositions, then at a later stage (in the general theory of relativity) we shall see that this "truth" is limited, and we shall consider the extent of its limitation.

II. SPACE AND TIME IN CLASSICAL MECHANICS

"The purpose of mechanics is to describe how bodies change their position in space with time." I should load my conscience with grave sins against the sacred spirit of lucidity were I to formulate the aims of mechanics in this way, without serious reflection and detailed explanations. Let us proceed to disclose these sins.

It is not clear what is to be understood here by "position" and "space." I stand at the window of a railway carriage which is travelling uniformly, and drop a stone on the embankment, without throwing it. Then, disregarding the influence of the air resistance, I see the stone descend in a straight line. A pedestrian who observes the misdeed from the footpath notices that the stone falls to earth in a parabolic curve. I now ask: Do the "positions" traversed by the stone lie "in reality" on a straight line or on a parabola? Moreover, what is meant here by motion "in space"? From the considerations of the previous section the answer is self-evident. In the first place, we entirely shun the vague word "space," of which, we must honestly acknowledge, we cannot form the slightest conception, and we replace it by "motion relative to a practically rigid body of reference." The positions relative to the body of reference (railway carriage or embankment) have already been defined in detail in the preceding section. If instead of "body of reference" we insert "system of co-ordinates," which is a useful idea for mathematical description, we are in a position to say: The stone traverses a straight line relative to a system of co-ordinates rigidly attached to the carriage, but relative to a system of co-ordinates rigidly attached to the ground (embankment) it describes a parabola. With the aid of this example it is clearly seen that there is no such thing as an independently existing trajectory (lit. "path-curve"), but only a trajectory relative to a particular body of reference.

In order to have a *complete* description of the motion, we must specify how the body alters its position *with time; i.e.* for every point on the trajectory it must be stated at what time the body is situated there. These data must be supplemented by such a definition of time that, in virtue of this definition, these time-values can be regarded essentially as magnitudes (results of measurements) capable of observation. If we take our stand on the ground of classical mechanics, we can satisfy this requirement for our illustration in the following manner. We imagine two clocks of identical construction; the man at the railway-carriage window is holding one of them, and the man on the footpath the other. Each of the observers determines the position on his own reference-body occupied by the stone at each tick of the clock he is holding in his hand. In this connection we have not taken account of the inaccuracy involved by the finiteness of the velocity of propagation of light. With this and with a second difficulty prevailing here we shall have to deal in detail later.

V. THE PRINCIPLE OF RELATIVITY (IN THE RESTRICTED SENSE)

In order to attain the greatest possible clearness, let us return to our example of the railway carriage supposed to be travelling uniformly. We call its motion a uniform translation ("uniform" because it is of constant velocity and direction, "translation" because although the carriage changes its position relative to the embankment yet it does not rotate in so doing). Let us imagine a raven flying through the air in such a manner that its motion, as observed from the embankment, is uniform and in a straight line. If we were to observe the flying raven from the moving railway carriage, we should find that the motion of the raven would be one of different velocity and direction, but that it would still be uniform and in a straight line. Expressed in an abstract manner we may say: If a mass m is moving uniformly in a straight line with respect to a co-ordinate system K, then it will also be moving uniformly and in a straight line relative to a second co-ordinate system K', provided that the latter is executing a uniform translatory motion with respect to K. In accordance with the discussion contained in the preceding section, it follows that:

If K is a Galileian co-ordinate system, then every other co-ordinate system K' is a Galileian one, when, in relation to K, it is in a condition of uniform motion of translation. Relative to K' the mechanical laws of Galilei-Newton hold good exactly as they do with respect to K.

We advance a step farther in our generalisation when we express the tenet thus: If, relative to K, K' is a uniformly moving co-ordinate system devoid of rotation, then natural phenomena run their course with respect to K' according to exactly the same general laws as with respect to K. This statement is called the *principle of relativity* (in the restricted sense).

As long as one was convinced that all natural phenomena were capable of representation with the help of classical mechanics, there was no need to doubt the validity of this principle of relativity. But in view of the more recent development of electrodynamics and optics it became more and more evident that classical mechanics affords an insufficient foundation for the physical description of all natural phenomena. At this juncture the question of the validity of the principle of relativity became ripe for discussion, and it did not appear impossible that the answer to this question might be in the negative.

Nevertheless, there are two general facts which at the outset speak very much in favour of the validity of the principle of relativity. Even though classical mechanics does not supply us with a sufficiently broad basis for the theoretical presentation of all physical phenomena, still we must grant it a considerable measure of "truth," since it supplies us with the actual motions of the heavenly bodies with a delicacy of detail little short of wonderful. The principle of relativity must therefore apply with great accuracy in the domain of *mechanics*. But that a principle of such broad generality should hold with such exactness in one domain of phenomena, and yet should be invalid for another, is *a priori* not very probable.

We now proceed to the second argument, to which, moreover, we shall return later. If the principle of relativity (in the restricted sense) does not hold, then the Galileian co-ordinate systems K, K', K'', etc., which are moving uniformly relative to each other, will not be *equivalent* for the description of natural phenomena. In this case we should be constrained to believe that natural laws are capable of being formulated in a particularly simple manner, and of course only on condition that, from amongst all possible Galileian co-ordinate systems, we should have chosen *one* (K_0) of a particular state of motion as our body of reference. We should then be justified (because of its merits for the description of natural phenomena) in calling this system "absolutely at rest," and all other Galileian systems K "in motion." If, for instance, our embankment were the system K_0, then our railway carriage would be a system K, relative to which less simple laws would hold than with respect to K_0. This diminished simplicity would be due to the fact that the carriage K would be in motion (*i.e.* "really") with respect to K_0. In the general laws of natural which have been formulated with reference to K, the magnitude and direction of the velocity of the carriage would necessarily play a part. We should expect, for instance, that the note emitted by an organ-pipe placed with its axis parallel to the direction of travel would be different from that emitted if the axis of the pipe were placed perpendicular to this direction. Now in virtue of its motion in an orbit round the sun, our earth is comparable with a railway carriage travelling with a velocity of about 30 kilometres per second. If the principle of relativity were not valid we should therefore expect that the direction of motion of the earth at any moment would enter into the laws of nature, and also that physical systems in their behaviour would be dependent on the orientation in space with respect to the earth. For owing to the alteration in direction of the velocity of rotation of the earth in the course of a year, the earth cannot be at rest relative to the hypothetical system K_0 throughout the whole year. However, the most careful observations have never revealed such anisotropic properties in terrestrial physical space, *i.e.* a physical non-equivalence of different directions. This is a very powerful argument in favour of the principle of relativity.

XVIII. SPECIAL AND GENERAL PRINCIPLE OF RELATIVITY

The basal principle, which was the pivot of all our previous considerations, was the *special* principle of relativity, *i.e.* the principle of the physical relativity of all *uniform* motion. Let us once more analyse its meaning carefully.

It was at all times clear that, from the point of view of the idea it conveys to us, every motion must only be considered as a relative motion. Returning to the illustration we have frequently used of the embankment and the railway carriage, we can express the fact of the motion here taking place in the following two forms, both of which are equally justifiable:

a.   The carriage is in motion relative to the embankment.

b.   The embankment is in motion relative to the carriage.

In (*a*) the embankment, in (*b*) the carriage, serves as the body of reference in our statement of the motion taking place. If it is simply a question of detecting or of describing the motion involved, it is in principle immaterial to what reference-body we refer the motion. As already mentioned, this is self-evident, but it must not be confused with the much more comprehensive statement called "the principle of relativity," which we have taken as the basis of our investigations.

The principle we have made use of not only maintains that we may equally well choose the carriage or the embankment as our reference-body for the description of any event (for this, too, is self-evident). Our principle rather asserts what follows: If we formulate the general laws of nature as they are obtained from experience, by making use of

a.   the embankment as reference-body,

b.   the railway carriage as reference-body,

then these general laws of nature (*e.g.* the laws of mechanics or the law of the propagation of light *in vacuo*) have exactly the same form in both cases. This can also be expressed as follows: For the *physical* description of natural processes, neither of the reference-bodies K, K' is unique (lit. "specially marked out") as compared with the other. Unlike the first, this latter statement need not of necessity hold *a priori*; it is not contained in the conceptions of "motion" and "referencebody" and derivable from them; only *experience* can decide as to its correctness or incorrectness.

Up to the present, however, we have by no means maintained the equivalence of all bodies of reference K in connection with the formulation of natural laws. Our course was more on the following lines. In the first place, we started out from the assumption that there exists a reference-body K, whose condition of motion is such that the Galileian law holds with respect to it: A particle left to itself and sufficiently far removed from all other particles moves uniformly in a straight line. With reference to K (Galileian reference-body) the laws of nature were to be as simple as possible. But in addition to K, all bodies of reference K' should be given preference in this sense, and they should be exactly equivalent to K for the formulation of natural laws, provided that they are in a state of *uniform rectilinear and non-rotary motion* with respect to K; all these bodies of reference are to be regarded as Galileian reference-bodies. The validity of the principle of relativity was assumed only for these reference-bodies, but not for others (*e.g.* those possessing motion of a different kind). In this sense we speak of the *special* principle of relativity, or special theory of relativity.

In contrast to this we wish to understand by the "general principle of relativity" the following statement: All bodies of reference K, K', etc., are equivalent for the description of natural phenomena (formulation of the general laws of nature), whatever may be their state of motion. But before proceeding farther, it ought to be pointed out that this formulation must be replaced later by a more abstract one, for reasons which will become evident at a later stage.

Since the introduction of the special principle of relativity has been justified, every intellect which strives after generalisation must feel the temptation to venture the step towards the general principle of relativity. But a simple and apparently quite reliable consideration seems to suggest that, for the present at any rate, there is little hope of success in such an attempt. Let us imagine ourselves transferred to our old friend the railway carriage, which is travelling at a uniform rate. As long as it is moving uniformly, the occupant of the carriage is not sensible of its motion, and it is for this reason that he can un-reluctantly interpret the facts of the case as indicating that the carriage is at rest, but the embankment in motion. Moreover, according to the special principle of relativity, this interpretation is quite justified also from a physical point of view.

CHAPTER 8
THE ATOM

All of the matter around us is made of atoms—the chemical building blocks of our world.

INTRODUCTION

T HE ATOMIC THEORY forms the basis of all chemistry. That all matter is made of atoms, and that the variety of atoms is systematized in the periodic table of the elements, is now well established, but these subjects provided a major focus of chemical research in the nineteenth century. The following three excerpts describe three pivotal discoveries regarding the nature of atoms and elements.

English naturalist John Dalton (1766–1844) made significant contributions to meteorology, the physics of gases (which he called "elastic fluids"), and color blindness (notably the condition of red-green color blindness from which he, himself, suffered, now known as Daltonism). However, his most famous contribution was the first modern articulation of the atomic theory of matter. The concept of atoms—the individual units that make up all of the matter around us—is thousands of years old. However, it was not until the 1803 publication of John Dalton's *New System of Chemical Philosophy* that the concept of a different type of atom for each chemical element was presented with a significant body of experimental support. Dalton came to this conclusions through his observations of the atmosphere: "Having been long accustomed to make meteorological observations, and to speculate on the nature and constitution of the atmosphere, it often struck me with wonder how a compound atmosphere, or a mixture of two or more elastic fluids, should constitute apparently a homogeneous mass …" The answer, he realized, lies in the particulate nature of each gas—the presence of sub-microscopic atoms. In these excerpts Dalton describes various common atmospheric gases that form from combinations of the elements hydrogen, oxygen, carbon ("charcoal"), and nitrogen ("azote"). Dalton's review of the chemical properties of water (just one of the many gases he examines in similar detail) provides a sense of the "new" chemistry of atoms.

New System of Chemical Philosophy
by John Dalton (1803)

CHAPTER II. ON THE CONSTITUTION OF BODIES

THERE ARE THREE distinctions in the kinds of bodies, or three states, which have more especially claimed the attention of philosophical chemists; namely, those which are marked by the terms *elastic fluids, liquids, and solids.* A very familiar instance is exhibited to us in water, of a body, which, in certain circumstances, is capable of assuming all the three states. In steam we recognise a perfectly elastic fluid, in water, a perfect liquid, and in ice a complete solid. These observations have tacitly led to the conclusion which seems universally adopted, that all bodies of sensible magnitude, whether liquid or solid, are constituted of a vast number of extremely small particles, or atoms of matter bound together by a force of attraction, which is more or less powerful according to circumstances, and which as it endeavours to prevent their separation, is very properly called in that view, *attraction of cohesion*; but as it collects them from a dispersed state (as from steam into water) it is called, *attraction of aggregation* or more simply, *affinity.* Whatever names it may go by, they still signify one and the same power. It is not my design to call in question this conclusion, which appears completely satisfactory; but to shew that we have hitherto made no use of it, and that the consequence of the neglect, has been a very obscure view of chemical agency, which is daily growing more so in proportion to the new lights attempted to be thrown upon it. ...

Whether the ultimate particles of a body, such as water, are all alike, that is, of the same figure, weight, &c. is a question of some importance. From what is known, we have no reason to apprehend a diversity in these particulars: if it does exist in water, it must equally exist in the elements constituting water, namely, hydrogen and oxygen. Now it is scarcely possible to conceive how the aggregates of dissimilar particles should be so uniformly the same. If some of the particles of water were heavier than others, if a parcel of the liquid on any occasion were constituted principally of these heavier particles, it must be supposed to affect the specific gravity of the mass, a circumstance not known. Similar observations may be made on other substances. Therefore we may conclude that *the ultimate particles of all homogeneous bodies are perfectly alike in weight, figure, etc.* In other words, every particle of water is like every other particle of water; every particle of hydrogen is like every other particle of hydrogen, etc. ...

CHAPTER III. ON CHEMICAL SYNTHESIS

When any body exists in the elastic [*i.e., gaseous*] state, its ultimate particles are separated from each other to a much greater distance than in any other stare; each particle occupies the centre of a comparatively large sphere, and supports its dignity by keeping all the rest, which by their gravity, or otherwise are disposed to encroach up it, at a respectful distance. When we attempt to conceive the *number* of particles in an atmosphere, it is somewhat like attempting to conceive the number of stars in the universe; we are confounded with the thought. But if we limit the subject, by taking a given volume of any gas, we seem persuaded that, let the divisions be ever so minute, the number of particles must be finite; just as in a given space of the universe, the number of stars and planets cannot be infinite.

Chemical analysis and synthesis go no farther than to the separation of particles one from another, and to their reunion. No new creation or destruction of matter is within the reach of chemical agency. We might as well attempt to introduce a new planet into the solar system, or to annihilate one already in existence, as to create or destroy a particle of hydrogen. All the changes we can produce, consist in separating particles that are in a state of cohesion or combination, and joining those that were previously at a distance.

In all chemical investigations, it has justly been considered an important object to ascertain the relative *weights* of the simples which constitute a compound. But unfortunately the enquiry has terminated here; whereas from the relative weights in the mass, the relative weights of the ultimate particles or atoms of the bodies might have been inferred, from which their number and weight in various other compounds would appear, in order to assist and to guide future investigations, and to correct their results. Now it is one great object of this work, to shew the importance and advantage of ascertaining *the relative weights of the ultimate particles, both of simple and compound bodies, the number of simple elementary particles which constitute one compound particle, and the number of less compound particles which enter into the formation of one more compound particle.* ...

From the application of these rules, to the chemical facts already well ascertained, we deduce the following conclusions; 1st. That water is a binary compound of hydrogen and oxygen, and the relative weights of the two elementary atoms are as 1:7, nearly; 2d. That ammonia is a binary compound of hydrogen and azote, and the relative weights of the two atoms are as 1:5, nearly; 3d. That nitrous gas is a binary compound of azote and oxygen, the atoms of which weigh 5 and 7 respectively; that nitric acid is a binary or ternary compound according as it is derived, and consists of one atom of azote and two of oxygen, together weighing 19; that nitrous oxide is a compound similar to nitric acid, and consists of one atom of oxygen and two of azote, weighing 17; that nitrous acid is a binary compound of nitric acid and nitrous gas, weighing 31; that oxynitric acid is a binary compound of nitric acid and oxygen, weighing 26; 4th. That carbonic oxide is a binary compound, consisting of one atom of charcoal, and one of oxygen, together weighing nearly 12; that carbonic acid is a ternary compound, (but sometimes binary) consisting of one atom of charcoal, and two of oxygen, weighing 19; etc. etc. In all these cases the weights are expressed in atoms of hydrogen, each of which is denoted by unity.

CHAPTER V. COMPOUNDS OF TWO ELEMENTS. OXYGEN WITH HYDROGEN
1. Water

This liquid, the most useful and abundant of any in nature, is now well known both by analytic and synthetic methods, to be a compound of the two elements, oxygen and hydrogen.

Canton has proved that water is in degree compressible. The expansive effect of heat on water has been already pointed out. The weight of a cubic foot of water is very near 1000 ounces avoirdupoise. This fluid is commonly taken as the standard for comparing the specific gravities of bodies, its weight being denoted by unity.

Distilled water is the purest; next to that, rain water; then river water; and, lastly, spring water. By purity in this place, is meant freedom from any foreign body in a state of solution; but in regard to transparency, and an agreeable taste, spring water generally excels the others. Pure water has the quality we call *soft*, spring and other impure water has the quality we call *hard*. Every one knows the great difference of waters in these respects; yet it is seldom that the hardest spring water contains so much as 1/1000th part of its weight of any foreign body in solution. The substances held in solution are usually carbonate and sulphate of lime.

Water usually contains about 2 percent of its bulk of common air. This air is originally forced into it by the pressure of the atmosphere; and can be expelled again no other way than by removing that pressure. This may be done by an air-pump; or it may in great part be effected by subjecting the water to ebullition, in which case steam takes the place of the incumbent air, and its pressure is found inadequate to restrain the dilatation of the air in the water, which of course makes its escape. But it is difficult to expel all the air by either of those operations. Air expelled from common spring water, after losing 5 or 10 percent of carbonic acid, consists of 38 percent of oxygen and 62 of azote.

Water is distinguished for entering into combination with other bodies. To some it unites in a small definite proportion, constituting a solid compound. This is the case in its combination with the fixed alkalies, lime, and with a great number of salts; the compounds are either dry powders or crystals. Such compounds have received the name of *hydrates*. But when the water is in excess, a different sort of combination seems to take place, which is called *solution*. In this case, the compound is *liquid* and transparent; as when common salt or sugar are dissolved in water. When any body is thus dissolved in water, it may be uniformly diffused through any larger quantity of that liquid, and seems to continue so, without manifesting any tendency to subside, as far as is known.

In 1781, the composition and decomposition of water were ascertained; the former by Watt and Cavendish, and the latter by Lavoisier and Meusnier. The first experiment on the composition of water on a large scale, was made by Monge, in 1783; he procured about 4 pounds of water, by the combustion of hydrogen gas, and noted the quantities of hydrogen and oxygen gas which had disappeared. The second experiment was made by Le Fevre de Gineau, in 1788; he obtained about 2 pounds of water in the same way. The third was made by Fourcroy, Vauquelin, and Seguin, in 1790, in which more than a pound of water was obtained. The general result was, that 85 parts by weight of oxygen unite to 15 of hydrogen to form 100 parts of water. —Experiments to ascertain the proportion of the elements arising from the decomposition of water, were made by Le Fevre de Gineau and by Lavoisier, by transmitting steam through a red hot tube containing a quantity of soft iron wire; the oxygen of the water combined with the iron, and the hydrogen was collected in gas. The same proportion, or 85 parts of oxygen and 15 of hydrogen, were found as in the composition.

The Dutch chemists, Dieman andTroostwyk, first succeeded in decomposing water by electricity, in 1789. The effect is now produced readily by galvanism. The composition of water is easily and elegantly shewn, by means of Volta's eudiometer, an instrument of the greatest importance in researches concerning elastic fluids. It consists of a strong graduated glass tube, into which a wire is hermetically sealed, or strongly cemented; another detached wire is pushed up the tube, nearly to meet the former, so that an electric spark or shock can be sent from one wire to the other through any portion of gas, or mixture of gases, confined by water or mercury. The end of the tube being immersed in a liquid, when an explosion takes place, no communication with the external air can arise; so that the change produced is capable of being ascertained.

The component parts of water being clearly established, it becomes of importance to determine with as much precision as possible, the relative weights of the two elements constituting that liquid. The mean results of analysis and synthesis, have given 85 parts of oxygen and 15 of hydrogen, which are generally adopted. In this estimate, I think, the quantity of hydrogen is overrated. There is an excellent memoir in the 53d vol. of the Annal. de Chemie, 1805, by Humboldt and Gay-Lussac, on the proportion of oxygen and hydrogen in water. They make it appear, that the quantity of aqueous vapour which elastic fluids usually contain, will so far influence the weight of hydrogen gas, as to change the more accurate result of Fourcroy, &c. of 85.7 oxygen and 14.3 hydrogen, to 87.4 oxygen and 12.6 hydrogen. Their reasoning appears to me perfectly satisfactory. The relation of these two numbers is that of 7 to 1 nearly. There is another consideration which seems to put this matter beyond doubt. In Volta's eudiometer, *two* measures of hydrogen require just *one* of oxygen to saturate them. Now, the accurate experiments of Cavendish and Lavoisier, have shewn that oxygen is nearly 14 times the weight of hydrogen; the exact coincidence of this with the conclusion above deduced, is a sufficient confirmation. If, however, any one chooses to adopt the common estimate of 85 to 15, then the relation of oxygen to hydrogen will be as 5.5 to 1; this would require the weight of oxygenous gas to be only 11 times the weight of hydrogen.

The absolute weights of oxygen and hydrogen in water being determined, the relative weights of their atoms may be investigated. As only *one* compound of oxygen and hydrogen is certainly known, it is agreeable to the 1st rule, page 214, that water should be concluded a *binary* compound; or, one atom of oxygen unites with one of hydrogen to form one of water. Hence, the relative weights of the atoms of oxygen and hydrogen are 7 to 1.

The above conclusion is strongly corroborated by other considerations. Whatever may be the proportions in which oxygen and hydrogen are mixed, whether 20 measures of oxygen to 2 of hydrogen, or 20 of hydrogen to 2 of oxygen, still when an electric spark is passed, water is formed by the union of 2 measures of hydrogen with 1 of oxygen, and the surplus gas is unchanged. Again, when water is decomposed by electricity, or by other agents, no other elements than oxygen and hydrogen are obtained. Besides, all the other compounds into which those two elements enter, will in the sequel be found to support the same conclusion.

After all, it must be allowed to be possible that water may be a ternary compound. In this case, if two atoms of hydrogen unite to one of oxygen, then an atom of oxygen must weigh 14 times as much as one of hydrogen; if two atoms of oxygen unite to one of hydrogen, then an atom of oxygen must weigh 3½ times one of hydrogen.

The Relation Between the Properties and Atomic Weights of the Elements

by Dimitri Ivanovich Mendeleev (1869)

DIMITRI IVANOVICH MENDELEEV (1834–1907) became a professor of chemistry at the University of St. Petersburg in 1867, shortly after his graduation with a doctorate degree from the same institution. At that time dozens of chemical elements had been discovered, and more were being added regularly through new techniques, including electrochemistry and spectroscopy. Hoping to bring some order to the subject of chemistry, Mendeleev began to systematize the known elements by arranging them in order of increasing weight. He quickly observed that familiar families of elements with similar properties, such as the alkali metals (lithium, sodium and potassium, designated Li, Na and K, respectively) and the alkaline earths (beryllium, magnesium and calcium, designated Be, Mg and Ca), could be arranged in vertical columns. He published his periodic table in 1869 with several gaps, which predicted the existence of elements unknown at the time. The subsequent discoveries of gallium (what Mendeleev called "ekaboron,") scandium, and germanium (elements 31, 21, and 32, respectively), all of which had properties matching Mendeleev's predictions, was seen as a great triumph for the model and ushered in widespread acceptance of the now-familiar periodic table of the elements.

* * *

...In undertaking to prepare a textbook called "Principles of Chemistry," I wished to establish some sort of system of simple bodies in which their distribution is not guided by chance, as might be thought instinctively, but by some sort of definite and exact principle. We previously saw that there was an almost complete absence of numerical relations for establishing a system of simple bodies, but in the end any system based on numbers which can be determined exactly will deserve preference over other systems which do not have numerical support, since the former leave little room for arbitrary choices. The numerical data for simple bodies are limited at the present time. If for some of them the physical properties are determined with certainty, yet this applies only to a very small number of the elementary bodies. For example, such properties as optical, or even electrical or magnetic, ones cannot in the end serve as a support for a system because one and the same body can show different values for these properties, depending on the state in which they occur. In this regard, it is enough to recall graphite and diamond, ordinary and red phosphorus, and oxygen and ozone. ... We know only one constant peculiar to an element, namely, the atomic weight. The size of the atomic weight, by the very essence of the matter, is a number which is not related to the state of division of the simple body but to the material part which is common to the simple body and all its compounds. The atomic weight belongs not to coal or the diamond, but to carbon. ... This is the reason I have chosen to base the system on the size of the atomic weights of the elements.

The first attempt which I made in this way was the following: I selected the bodies with the lowest atomic weights and arranged them in the order of the size of their atomic weights. This showed that there existed a period in the properties of the simple bodies, and even in terms of their atomicity the elements followed each other in the order of arithmetic succession of the size of their atoms:

Li = 7; Be = 9.4; B = 11;   C = 12; N = 14; O = 16; F = 19
Na = 23; Mg = 24; Al = 27.4;   Si = 28; P = 31; S = 32; Cl = 35.3
K = 39; Ca = 40; ...   Ti = 50; V = 51 ...

In the arrangement of elements with atoms greater than 100, we meet an entirely analogous continuous order:
Ag = 108; Cd = 112; Ur = 116; Sn = 118; Sb = 122; Te = 128; I = 127.

It has been shown that Li, Na, K, and Ag are related to each other, as are C, Si, Ti, Sn, or as are N, P, V, Sb, etc. This at once raises the question whether the properties of the elements are expressed by their atomic weights and whether a system can be based on them. An attempt at such a system follows.

In the assumed system, the atomic weight of the element, unique to it, serves as a basis for determining the place of the element. Comparison of the groups of simple bodies known up to now according to the weights of their atoms leads to the conclusion that the distribution of the elements according to their atomic weights does not disturb the natural similarities which exist between the elements but, on the contrary, shows them directly. ...

All the comparisons which I have made in this direction lead me to conclude that *the size of the atomic weight determines the nature of the elements,* just as the weight of the molecules determines the properties and many of the reactions of complex bodies. If this conclusion is confirmed by further applications of this approach to the study of the elements, then we are near an epoch in understanding, the existing differences and the reasons for the similarity of elementary bodies.

I think that the law established by me does not run counter to the general direction of natural science, and that until now it has not been demonstrated, although already there have been hints of it. Henceforth, it seems to me, there will be a new interest in determining atomic weights, in discovering new elementary bodies, and in finding new analogies between them.

I now present one of many possible systems of elements based on their atomic weights. It serves only as an attempt to express those results which can be obtained in this way. I myself see that this attempt is not final, but it seems to me that it clearly expresses the applicability of my assumptions to all combinations of elements whose atoms are known with certainty. In this I have also wished to establish a general system of the elements. ...

In conclusion, I consider it advisable to recapitulate the results of the above work.

1. Elements arranged according to the size of their atomic weights show clear *periodic* properties.

2. Elements which are similar in chemical function either have atomic weights which lie close together (like Pt, Ir, Os) or show a uniform increase in atomic weight (like K, Rb, Cs). ...

3. Comparisons of the elements or their groups in terms of size of their atomic weights establish their so-called "atomicity" and, to some extent, differences in chemical character, a fact which is clearly evident in the group Li, Be, B, C, N, O, F, and is repeated in the other groups.

4. The simple bodies which are most widely distributed in nature have small atomic weights, and all the elements which have small atomic weights are characterized by the specificity of their properties. They are therefore the typical elements. Hydrogen, as the lightest element, is in justice chosen as typical of itself.

5. The *size* of the atomic weight determines the character of the element, just as the size of the molecule determines the properties of the complex body, and so, when we study compounds, we should consider not only the properties and amounts of the elements, not

only the reactions, but also the weight of the atoms. Thus, for example, compounds of S and Te, CI and I, etc., although showing resemblances, also very clearly show differences.

6. We should still expect to discover many *unknown* simple bodies; for example, those similar to Al and Si, elements with atomic weights of 65 to 75.

7. Some *analogies* of the elements are discovered from the size of the weights of their atoms. Thus uranium is shown to be analogous to boron and aluminum, a fact which is also justified when their compounds are compared.

The purpose of my paper will be entirely attained if I succeed in turning the attention of investigators to the same relationships in the size of the atomic weights of nonsimilar elements, which have, as far as I know, been almost entirely neglected until now. Assuming that in problems of this nature lies the solution of one of the most important questions of our science, I myself, as my time will permit, will turn to a comparative study of lithium, beryllium, and boron. ...

And now, in order to clarify the matter further, I wish to draw some conclusions as to the chemical and physical properties of those elements which have not been placed in the system and which are still undiscovered but whose discovery is very probable. I think that until now we have not had any chance to foresee the absence of these or other elements, because we have had no order for their arrangement, and even less have we had occasion to predict the properties of such elements. An established system is limited by its order of known or discovered elements. With the periodic and atomic relations now shown to exist between all the atoms and the properties of their elements, we see the possibility not only of noting the absence of some of them but even of determining, and with great assurance and certainty, the properties of these as yet unknown elements; it is possible to predict their atomic weight, density in the free state or in the form of oxides, acidity or basicity, degree of oxidation, and ability to be reduced and to form double salts and to describe the properties of the metalloorganic compounds and chlorides of the given element; it is even possible also to describe the properties of some compounds of these unknown elements in still greater detail. Although at the present time it is not possible to say when one of these bodies which I have predicted will be discovered, yet the opportunity exists for finally convincing myself and other chemists of the truth of those hypotheses which lie at the base of the system I have drawn up. Personally, for me these assumptions have become so strong that, as in the case of indium, there is justification for the ideas which are based on the periodic law which lies at the base of all this study.

Among the ordinary elements, the *lack* of a number of *analogues of boron and aluminum* is very striking, that is, in group III, and it is certain that we lack an element of this group immediately following aluminum; this must be found in the even, or second, series, immediately after potassium and calcium. Since the atomic weights of these latter are near 40, and since then in this row the element of group IV, titanium, Ti = 50, follows, then the atomic weight of the missing element should be nearly 45. Since this element belongs to an even series, it should have more basic properties than the lower elements of group III, boron or aluminum, that is, its oxide, R_2O_3, should be a stronger base. An indication of this is that the oxide of titanium, TiO_2, with the properties of a very weak acid, also shows many signs of being clearly basic. On the basis of these properties, the oxide of the metal should still be weak, like the weakly basic properties of titanium dioxide; compared to aluminum, this oxide should have a more strongly basic character, and therefore, probably, it should not decompose water, and it should combine with acids and alkalis to form simple salts; ammonia will not dissolve it, but perhaps the hydrate will dissolve weakly in potassium hydroxide, although the latter is doubtful because the element belongs to the even series and to a group of elements whose oxides contain a small amount of oxygen. I have decided to give this element the preliminary name of *ekaboron*, deriving the name from this, that it follows boron as the first element of the even group, and the syllable *eka* comes from the Sanskrit word meaning "one." Eb = 45. Ekaboron should be a metal with an atomic volume of about 15, because in the elements of the second series, and in all the even series, the atomic volume falls quickly as we go from the first group to the following ones. Actually, the volume of potassium is nearly 50, calcium nearly 25, titanium and vanadium nearly 9, and chromium, molybdenum, and iron nearly 7; thus the specific gravity of the metal should be close to 3.0, since its atomic weight = 45. The metal will be nonvolatile, because all the metals in the even series of all the groups (except group I) are nonvolatile; hence it can hardly be discovered by the ordinary method of spectrum analysis. It should not decompose water at ordinary temperature, but at somewhat raised temperatures it should decompose it, as do many other metals of this series which form basic oxides. Finally, it will dissolve in acids. Its chloride $EbCl_3$ (perhaps Eb_2Cl_6), should be a volatile substance but a salt, since it corresponds to a basic oxide. Water will act on it as it does on the chlorides of calcium and magnesium, that is, ekaboron chloride will be a hygroscopic body and will be able to evolve hydrogen chloride without having the character of a hydrochloride. Since the volume of calcium chloride = 49 and that of titanium chloride = 109, the volume of ekaboron chloride should be close to 78, and therefore its specific gravity will probably be about 2.0. Ekaboron oxide, Eb_2O_3, should be a nonvolatile substance and probably should not fuse; it should be insoluble in water, because even calcium oxide is very slightly soluble in water, but it will probably dissolve in acid. Its specific volume should be about 39, because in the series potassium oxide has a volume of 35, CaO = 18, TIO = 20, and CrO_3 = 36; that is, considered on the basis of a content of one atom of oxygen, the volume quickly falls to the right, thus, for potassium = 35 for calcium = 18, for titanium = 10, for chromium = 12, and therefore the volume for ekaboron oxide containing one atom of oxygen should be nearly 13, and so the formula Eb_2G_3 should correspond to a volume of about 39, and therefore anhydrous ekaboron oxide will have a specific gravity close to 3.5. Since it is a sufficiently strong base, this oxide should show little tendency to form alums, although it will probably give alum-forming compounds, that is, double salts with potassium sulfate. Finally, ekaboron will not form metalloorganic compounds, since it is one of the metals of an even series. Judging by the data now known for the elements which accompany cerium, none of them belong in the place which is assigned to ekaboron, so that this metal is certainly not one of the members of the cerium complex which is now known.

On a New Radioactive Substance Contained in Pitchblende[1]

by Marie Sklowdowska Curie (1903)

THE DISCOVERY OF RADIOACTIVITY by Henri Becquerel (1852–1908) in 1896 pointed to the possible existence of previously unknown radioactive chemical elements. The discoveries of polonium and radium (elements 84 and 88, respectively) were made within a few years by Polish-born Marie Sklowdowska Curie (1867–1934), who won Nobel Prizes in both Physics (in 1903, with her husband Pierre and Becquerel) and Chemistry (in 1911) for her pioneering studies. Though a brilliant student in her native Poland, she did not begin her scientific studies until she moved to Paris. There she met and married Pierre Curie (1859–1906), a faculty member at the Sorbonne who collaborated on some of her most famous experiments. The following excerpts come from an English translation of Marie Curie's 1903 doctoral thesis. She describes the laborious process of extracting minute amounts of the radioactive elements from the valuable uranium ore, pitchblende. In spite of her great scientific advances, she was refused admission to the exclusively male French Academy of Sciences. Her lifelong exposure to high doses of radiation likely contributed to her death from cancer in 1934.

* * *

CHAPTER II. METHOD OF RESEARCH

The results of the investigation of radio-active minerals, announced in the preceding chapter, led M. Curie and myself to endeavour to extract a new radio-active body from pitchblende. Our method of procedure could only be based on radio-activity, as we know of no other property of the hypothetical substance. The following is the method pursued for a research based on radio-activity. The radio-activity of this compound is determined, and a chemical decomposition of this compound is effected; the radio-activity of all the products

---

1   English translation from *Chemical News*, volume 88 (19030.

obtained is determined, having regard to the proportion in which the radio-active substance is distributed among them. In this way, an indication is obtained, which may to a certain extent be compared to that which spectrum analysis furnishes. In order to obtain comparable figures, the activity of the substances must be determined in the solid form well dried.

POLONIUM, RADIUM, ACTINIUM

The analysis of pitchblende with the help of the method just explained, led us to the discovery in this mineral of two strongly radioactive substances, chemically dissimilar:— Polonium, discovered by ourselves, and radium, which we discovered in conjunction with M. Bémont.

Polonium from the analytical point of view, is analogous to bismuth, and separates out with the latter. By one of the following methods of fractionating, bismuth products are obtained increasingly rich in polonium:—

1. Sublimation of the sulphides *in vacuo*; the active sulphide is much more volatile than bismuth sulphide.

2. Precipitation of solutions of the nitrate by water; the precipitate of the basic nitrate is much more active than the salt which remains in solution.

3. Precipitation by sulphuretted hydrogen of a hydrochloric acid solution, strongly acid; the precipitated sulphides are considerably more active than the salt which remains in solution.

Radium is a substance which accompanies the barium obtained from pitchblende; it resembles barium in its reactions, and is separated from it by difference of solubility of the chlorides in water, in dilute alcohol, or in water acidified with hydrochloric acid. We effect the separation of the chlorides of barium and radium by subjecting the mixture to fractional crystallisation, radium chloride being less soluble than that of barium.

A third strongly radio-active body has been identified in pitchblende by Mo Debierne, who gave it the name of *actinium*. Actinium accompanies certain members of the iron group contained in pitchblende; it appears in particular allied to thorium, from which it has not yet been found possible to separate it. The extraction of actinium from pitchblende is a very difficult operation, the separations being as a rule incomplete.

All three of the new radio-active bodies occur in quite infinitesimal amount in pitchblende. In order to obtain them in a more concentrated condition, we were obliged to treat several tons of residue of the ore of uranium. The rough treatment was carried out in the factory; and this was followed by processes of purification and concentration. We thus succeeded in extracting from thousands of kilogrammes of crude material a few decigrammes of products which were exceedingly active as compared with the ore from which they were obtained. It is obvious that this process is long, arduous, and costly. ...

EXTRACTION OF THE NEW RADIO-ACTIVE SUBSTANCES

The first stage of the operation consists in extracting barium with radium from the ores of uranium, also bismuth with polonium and the rare earths containing actinium from the same. These three primary products having been obtained, the next step is in each case to endeavour to isolate the new radio-active body. This second part of the treatment consists of a process of fractionation. The difficulty of finding a very perfect means of separating closely allied elements is well known; methods of fractionation are therefore quite suitable. Besides this, when a mere trace of one element is mixed with another element, no method of complete separation could be applied to the mixture, even allowing that such a method was known; in fact, one would run the risk of losing the trace of the material to be separated.

The particular object of my work has been the isolation of radium and polonium. After working for several years, I have so far only succeeded in obtaining the former.

Pitchblende is an expensive ore, and we have given up the treatment of it in large quantities. In Europe the extraction of this ore is carried out in the mine, of Joachimsthal, in Bohemia. The crushed ore is roasted with carbonate of soda, and the resulting material washed, first with warm water and then with dilute sulphuric acid. The solution contains the uranium, which gives pitchblende its value. The insoluble residue is rejected. This residue contains radio-active substances; its activity is four and a half times that of metallic uranium. The Austrian Government, to whom the mine belongs, presented us with a ton of this residue for our research, and authorised the mine to give us several tons more of the material.

It was not very easy to apply the methods of the laboratory to the preliminary treatment of the residue in the factory. M. Debierne investigated this question, and organised the treatment in the factory. The most important point of his method is the conversion of the sulphates into carbonate by boiling the material, with a concentrated solution of sodium: carbonate. This method avoids the necessity of fusing with sodium carbonate.

The residue chiefly contains the sulphates of lead and calcium, silica, alumina, and iron oxide. In addition nearly all the metals are found in greater or smaller amount (copper, bismuth, zinc, cobalt, manganese, nickel, vanadium, antimony, thallium, rare earths, niobium, tantalum, arsenic, barium, &c). Radium is found in this mixture as sulphate, and is the least soluble sulphate in it. In order to dissolve it, it is necessary to remove the sulphuric acid as far as possible. To do this, the residue is first treated with a boiling concentrated soda solution. The sulphuric acid combined with the lead, aluminium, and calcium passes, for the most part, into solution as sulphate of sodium, which is removed by repeatedly washing with water. The alkaline solution removes at the same time lead, silicon, and aluminium. The insoluble portion is attacked by ordinary hydrochloric acid. This operation completely disintegrates the material, and dissolves most of it. Polonium and actinium may be obtained front this solution; the former is precipitated by sulphuretted hydrogen, the: latter is found in the hydrates precipitated by ammonia in the solution separated from the sulphides and oxidised. Radium remains in the insoluble portion. This portion is washed with water, and then treated with a boiling concentrated solution of carbonate of soda. This operation completes the transformation of the sulphates of barium and radium into carbonates. The material is then thoroughly washed with water, and then treated with dilute hydrochloric acid, quite free from sulphuric acid. The solution contains radium as well as polonium and actinium. It is filtered and precipitated with sulphuric acid. In this way the crude sulphates of barium containing radium and calcium, of lead, and of iron, and of a trace of actinium are obtained. The solution still contains a little actinium and polonium, which may be separated out as in the case of the first-hydrochloric acid solution.

From one ton of residue 10 to 20 kilograms of crude sulphates are obtained, the activity of which is from thirty to sixty times as great, as that of metallic uranium. They must now be purified. For this purpose they are boiled with sodium carbonate and transformed into the chlorides. The solution is treated with sulphuretted hydrogen, which gives a small quantity of active sulphides containing polonium. The solution is filtered, oxidised by means of chlorine, and precipitated with pure ammonia. The precipitated hydrates and oxides are very active, and the activity is due to actinium. The filtered solution is precipitated with sodium carbonate. The precipitated carbonates of the alkaline earths are washed and converted into chlorides. These chlorides are evaporated to dryness, and washed with pure concentrated hydrochloric acid. Calcium chloride dissolves almost entirely, whilst the chloride of barium and radium remains insoluble. Thus, from one ton of the original material about 8 kilograms of barium and radium chloride are obtained, of which the activity is about sixty times that of metallic uranium. The chloride is now ready for fractionation. …

CONCLUSIONS

I will define, in conclusion, the part I have personally taken in the researches upon radio-active bodies.

I have investigated the radio-activity of uranium compounds. I have examined other bodies for the existence of radio-activity, and found the property to be possessed by thorium compounds. I have made clear the atomic character of the radio-activity of the compounds of uranium and thorium,

I have conducted a research upon radio-active substances other than uranium and thorium. To this end I investigated a large number of substances by an accurate electrometric method, and I discovered that certain minerals possess activity which is not to be accounted for by their content of uranium and thorium.

From this I concluded that these minerals must contain a radio-active body different from uranium and thorium, and more strongly radio-active than the latter metals.

In conjunction with M. Curie, and subsequently MM. Curie and Bémont, I was able to extract from pitchblende two strongly radio-active bodies—polonium and radium.

I have been continuously engaged upon the chemical examination and preparation of these substances. I effected the fractionations necessary to the concentration of radium and I succeeded in isolating pure radium chloride. Concurrently with this work, I made several atomic weight determinations with a very small quantity of material, and was finally able to determine the atomic weight of radium with a very fair degree of accuracy. The work has proved *that radium is a new chemical element.* Thus the new method of investigating new chemical elements, established by M. Curie and myself, based upon radio-activity, is fully justified.

I have investigated the law of absorption of polonium rays, and of the absorbable rays of radium, and have demonstrated that this law of absorption is peculiar and different from the known laws of other radiations.

I have investigated the variation of activity of radium salts, the effect of solution and of heating, and the renewal of activity with time, after solution or after heating.

In conjunction with M. Curie, I have examined different effects produced by the new radio-active substances (electric, photographic, fluorescent, luminous colourations, &c).

In conjunction with M. Curie, I have established the fact that radium gives rise to rays charged with negative electricity.

Our researches upon the new radio-active bodies have given rise to a scientific movement, and have been the starting-point of numerous researches in connection with new radio-active substances, and with the investigation of the radiation of the known radio-active bodies.

CHAPTER 9
QUANTUM MECHANICS

At the subatomic scale everything is quantized. Any measurement at that scale significantly alters the object being measured.

INTRODUCTION

E RWIN SCHRODINGER (1887–1961) was born in Vienna and took his doctorate in theoretical physics there, at a time when the city was both the capitol of the Austro-Hungarian Empire and one of the great cultural centers of Europe. He eventually moved on to the chair of theoretical physics in Berlin, but used an acquaintance with Irish president Eamon de Valera to move to Dublin at the beginning of World War II.

During the 1920s a relatively small group of scientists, mostly young men, mostly German, developed the new science of quantum mechanics to explain experimental information that was coming in on the behavior of matter at the atomic and sub-atomic level (see the readings for Neils Bohr and J. J. Thomson). Werner Hiesenberg and Schrodinger developed seemingly different theories, but it later turned out that they were simply different mathematical representations of the same thing. Schrodinger's "wave mechanics" requires less mathematical rigor, and is therefore what is usually taught these days. In his Nobel lecture, he departed from the usual custom of describing his work in technical detail to give a tutorial on wave mechanics and top introduce the problems (like wave-particle duality) that still bother people today.

The Fundamental Idea of Wave Mechanics
by Erwin Schrödinger
Nobel Lecture, December 12, 1933

ON PASSING THROUGH an optical instrument, such as a telescope or a camera lens, a ray of light is subjected to a change in direction at each refracting or reflecting surface. ...

In different media, light propagates with different velocities, and the radiation path gives the appearance as if the light must arrive at its destination *as quickly as possible.* ... This is the famous Fermat *principle of the shortest light time*, which in a marvelous manner determines the entire fate of a ray of light by a single statement and also includes the more general case, when the nature of the medium varies not suddenly at individual surfaces, but gradually from place to place.

According to the wave theory of light, the light rays, strictly speaking, have only fictitious significance. They are not the physical paths of some particles of light, but are a mathematical device, the so-called orthogonal trajectories of wave surfaces, imaginary guide lines as it were, which point in the direction normal to the wave surface in which the latter advances. It is surprising that a general principle as important as Fermat's relates directly to these mathematical guidelines, and not to the wave surfaces, and one might be inclined for this reason to consider it a mere mathematical curiosity. Far from it. It becomes properly understandable only from the point of view of wave theory and ceases to be a divine miracle. From the wave point of view, the so-called *curvature* of the light ray is far more readily understandable as a *swerving* of the wave surface, which must obviously occur when neighbouring parts of a wave surface advance at different speeds; in exactly the same manner as a company of soldiers marching forward will carry out the order "right incline" by the men taking steps of varying lengths, the right-wing man the smallest, and the left-wing man the longest.

The Fermat principle thus appears to be the trivial quintessence of the wave theory. It was therefore a memorable occasion when Hamilton made the discovery that the true movement of mass points in a field of forces (e.g. of a planet on its orbit around the sun or of a stone thrown in the gravitational field of the earth) is also governed by a very similar general principle, which carries and has made famous the name of its discoverer since then. Admittedly, the Hamilton principle does not say exactly that the mass point chooses the quickest way, but it does say something so similar—the analogy with the principle of the shortest travelling time of light is so close, that one was faced with a puzzle. It seemed as if Nature had realized one and the same law twice by entirely different means: first in the case of light, by means of a fairly obvious play of rays; and again in the case of the mass points, which was anything but obvious, unless somehow wave nature were to be attributed to them also. And this, it seemed impossible to do. Because the "mass points" on which the laws of mechanics had really been confirmed experimentally at that time were only the large, visible, sometimes very large bodies, the planets, for which a thing like "wave nature" appeared to be out of the question.

The smallest, elementary components of matter which we today, much more specifically, call "mass points," were purely hypothetical at the time. It was only after the discovery of radioactivity that constant refinements of methods of measurement permitted the properties of these particles to be studied in detail, and now permit the paths of such particles to be photographed and to be measured very exactly. As far as the measurements extend they confirm that the same mechanical laws are valid for particles as for large bodies, planets, etc. However, it was found that neither the molecule nor the individual atom can be considered as the "ultimate component": but even the atom is a system of highly complex structure.

Images are formed in our minds of the structure of atoms *consisting of* particles, images which seem to have a certain similarity with the planetary system. It was only natural that the attempt should at first be made to consider as valid the same laws of motion that had proved themselves so amazingly satisfactory on a large scale. In other words, Hamilton's mechanics, which, as I said above, culminates in the Hamilton principle, were applied also to the "inner life" of the atom. That there is a very close analogy between Hamilton's principle and Fermat's optical principle had meanwhile become all but forgotten. If it was remembered, it was considered to be nothing more than a curious trait of the mathematical theory.

Now, it is very difficult, without further going into details, to convey a proper conception of the success or failure of these classical-mechanical images of the atom. On the one hand, Hamilton's principle in particular proved to be the most faithful and reliable guide, which was simply indispensable; on the other hand one had to suffer, to do justice to the facts, the rough interference of entirely new incomprehensible postulates, of the so-called quantum conditions and quantum postulates. Strident disharmony in the symphony of classical mechanics—yet strangely familiar—played as it were on the same instrument. …

The situation was fairly desperate. Had the old mechanics failed completely, it would not have been so bad. The way would then have been free to the development of a new system of mechanics. As it was, one was faced with the difficult task of saving the *soul* of the old system, whose inspiration clearly held sway in this microcosm, while at the same time flattering it as it were into accepting the quantum conditions not as gross interference but as issuing from its own innermost essence.

The way out lay just in the possibility, already indicated above, of attributing to the Hamilton principle, also, the operation of a wave mechanism on which the point-mechanical processes are essentially based, just as one had long become accustomed to doing in the case of phenomena relating to light and of the Fermat principle which governs them. Admittedly, the individual path of a mass point loses its proper physical significance and becomes as fictitious as the individual isolated ray of light. The essence of the theory, the minimum principle, however, remains not only intact, but reveals its true and simple meaning only under the wave-like aspect, as already explained. Strictly speaking, the new theory is in fact not *new*, it is a completely organic development, one might almost be tempted to say a more elaborate exposition, of the old theory.

How was it then that this new more "elaborate" exposition led to notably different results; what enabled it, when applied to the atom, to obviate difficulties which the old theory could not solve? What enabled it to render gross interference acceptable or even to make it its own?

Again, these matters can best be illustrated by analogy with optics… The so-called refraction and interference phenomena of light can only be understood if we trace the wave process in detail because what matters is not only the eventual destination of the wave, but also whether at a given moment it arrives there with a wave peak or a wave trough. In the older, coarser experimental arrangements, these phenomena occurred as small details only and escaped observation. Once they were noticed and were interpreted correctly, by means of waves, it was easy to devise experiments in which the wave nature of light finds expression not only in small details, but on a very large scale in the entire character of the phenomenon.

[An] example is the shadow of an opaque object cast on a screen by a small point light source. In order to construct the shape of the shadow, each light ray must be traced and it must be established whether or not the opaque object prevents it from reaching the screen. The *margin* of the shadow is formed by those light rays which only just brush past the edge of the body. Experience has shown that the shadow margin is not absolutely sharp even with a point-shaped light source and a sharply defined shadow-casting object. The reason for this is the same as in the first example. The wave front is as it were bisected by the body (cf. Fig. 6) and the traces of this injury result in blurring of the margin of the shadow which would be incomprehensible if the individual light rays were independent entities advancing independently of one another without reference to their neighbours.

This phenomenon—which is also called diffraction—is not as a rule very noticeable with large bodies. But if the shadow-casting body is very small at least in one dimension, diffraction finds expression firstly in that no proper shadow is formed at all, and secondly—much more strikingly—in that the small body itself becomes as it were its own source of light and radiates light in all directions (preferentially to be sure, at small angles relative to the incident light). All of you are undoubtedly familiar with the so-called "motes of dust" in a light beam falling into a dark room. Fine blades of grass and spiders' webs on the crest of a hill with the sun behind it, or the errant locks of hair of a man standing with the sun behind often light up mysteriously by diffracted light, and the visibility of smoke and mist is based on it. It comes not really from the body itself, but from its immediate surroundings, an area in which it causes considerable interference with the incident wave fronts. It is interesting, and important for what follows, to observe that the area of interference always and in every direction has at least the extent of one or a few wavelengths, no matter how small the disturbing particle may be. Once again, therefore, we observe a close relationship between the phenomenon of diffraction and wavelength.

Let us return from optics to mechanics and explore the analogy to its fullest extent. In optics the old system of mechanics corresponds to intellectually operating with isolated mutually independent light rays. The new undulatory mechanics corresponds to the wave theory of light. What is gained by changing from the old view to the new is that the diffraction phenomena can be accommodated or, better expressed, what is gained is something that is strictly analogous to the diffraction phenomena of light and which on the whole must be very unimportant, otherwise the old view of mechanics would not have given full satisfaction so long. It is, however, easy to surmise that the neglected phenomenon may in some circumstances make itself very much felt, will entirely dominate the mechanical process, and will face the old system with insoluble riddles, if *the entire mechanical system is comparable in extent with the wavelengths of the "waves of matter"* which play the same part in mechanical processes as that played by the light waves in optical processes.

This is the reason why in these minute systems, the atoms, the old view was bound to fail, which though remaining intact as a close approximation for gross mechanical processes, but is no longer adequate for the delicate interplay in areas of the order of magnitude of one or a few wavelengths. It was astounding to observe the manner in which all those strange additional requirements developed spontaneously from the new undulatory view, whereas they had to be forced upon the old view to adapt them to the inner life of the atom and to provide some explanation of the observed facts.

Thus, the salient point of the whole matter is that the diameters of the atoms and the wavelength of the hypothetical material waves are of approximately the same order of magnitude. …

I have tried to place before you the fundamental idea of the wave theory of matter in the simplest possible form. I must admit now that in my desire not to tangle the ideas from the very beginning, I have painted the lily. Not as regards the high degree to which all sufficiently, carefully drawn conclusions are confirmed by experience, but with regard to the conceptual ease and simplicity with which the conclusions are reached. I am not speaking here of the mathematical difficulties, which always turn out to be trivial in the end, but of the conceptual difficulties. It is, of course, easy to say that we turn from the concept of a *curved path* to a system of wave surfaces normal to it. The wave surfaces, however, even if we consider only small parts of them include at least a narrow *bundle* of possible curved paths, to all of which they stand in the same relationship. According to the old view, but not according to the new, one of them in each concrete individual case is distinguished from all the others which are "only possible," as that "really travelled." We are faced here with the full force of the logical opposition between an

either – or (point mechanics)

and a

both – and (wave mechanics)

This would not matter much, if the old system were to be dropped entirely and to be *replaced* by the new. Unfortunately, this is not the case. From the point of view of wave mechanics, the infinite array of possible point paths would be merely fictitious, none of them would have the prerogative over the others of being that really travelled in an individual case. I have, however, already mentioned that we have yet really observed such individual particle paths in some cases. The wave theory can represent this, either not at all or only very imperfectly. We find it confoundedly difficult to interpret the traces we see as nothing more than narrow bundles of equally possible paths between which the wave surfaces establish cross-connections. Yet, these cross-connections are necessary for an understanding of the diffraction and interference phenomena which can be demonstrated for the same particle with the same plausibility—and that on a large scale, not just as a consequence of the theoretical ideas about the interior of the atom, which we mentioned earlier. Conditions are admittedly such that we can always manage to make do in each concrete individual case without the two different aspects leading to different expectations as to the result of certain experiments. We cannot, however, manage to make do with such old, familiar, and seemingly indispensible terms as "real" or "only possible"; we are never in a position to say what really *is* or what really *happens*, but we can only say what will be *observed* in any concrete individual case. Will we have to be permanently satisfied with this …? On principle, yes. On principle, there is nothing new in the postulate that in the end exact science should aim at nothing more than the description of what can really be observed. The question is only whether from now on we shall have to refrain from tying description to a clear hypothesis about the real nature of the world. There are many who wish to pronounce such abdication even today. But I believe that this means making things a little too easy for oneself.

I would define the present state of our knowledge as follows. The ray or the particle path corresponds to a *longitudinal* relationship of the propagation process (i.e. *in the direction* of propagation), the wave surface on the other hand to a *transversal* relationship (i.e. *normal* to it). *Both* relationships are without doubt real; one is proved by photographed particle paths, the other by interference experiments. To combine both in a uniform system has proved impossible so far. Only in extreme cases does either the transversal, shell-shaped or the radial, longitudinal relationship predominate to such an extent that we *think* we can make do with the wave theory alone or with the particle theory alone.

CHAPTER 10
THE CHEMICAL BOND

Atoms bind together in chemical reactions by the rearrangement of electrons.

INTRODUCTION

FOR HUNDREDS OF years, the central focus of the chemical sciences has been the study of chemical reactions, by which atoms are rearranged to form new products from reactants. The two excerpts in this section chronicle attempts to separate atomic mixtures into their constituent elements.

Swedish chemist Carl Wilhelm Scheele (1742–1786) discovered numerous new substances during his relatively brief career, including the gaseous elements oxygen and chlorine, and numerous important organic compounds. In his most cited work, Scheele performed a series of experiments on air and demonstrated that it is composed primarily of two distinct gases (or "elastic fluids"). The more abundant of these gases ("vitiated air," or nitrogen) comprises 80 percent of the atmosphere and does not support combustion. The other gas ("fire air," or oxygen), representing approximately 20 percent of the atmosphere, is essential for burning. These excerpts from Scheele's *Chemical Treatise on Air and Fire* outline some of the many experiments he performed to separate and characterize nitrogen and oxygen.

Chemical Treatise on Air and Fire
by Carl Wilhelm Scheele (1777)

PREFACE

THE INVESTIGATION OF the air is an important object of chemistry at the present time. This elastic fluid is endowed, too, with so many special properties that it can furnish material enough for new discoveries to anyone who takes such experiments in hand. Fire, this product of chemistry which is so wonderful, shows us that it cannot be generated without air; and do I, indeed, err if I have undertaken to adduce proofs in this treatise, which is only to be looked upon as an attempt towards a chemical theory of fire, that an air existent in our atmosphere is to be regarded as a true constituent of fire and consequently contributes materially to flame; wherefore I have named this air fire air? Certainly, I shall not be so bold as to press my readers to believe this. No, there are clear experiments which tell in my favour, experiments that I have made on more than a single occasion, and in which, if I do not mistake, I have sufficiently nearly attained my object of learning to understand, fire as clearly as possible. And this is the reward which I have obtained for my labour, and which, has occasioned me a true satisfaction that I cannot possibly retain for myself alone. This, and no other, is the reason why I make this work known to my readers. I had already carried out the greater part of these experiments when I obtained sight of Priestley's elegant observations. ...

1. It is the object and chief business of chemistry skillfully to separate substances into their constituents, to discover their properties, and to compound them in different ways.

How difficult it is, however, to carry out such operations with the greatest accuracy can only be unknown to one who either has never undertaken this occupation, or at least has not done so with sufficient attention.

2. Hitherto chemical investigators are not agreed as to how many elements or fundamental materials compose all substances. This is indeed one of the most difficult problems. Some hold that there remains no further hope of searching out the elements of substances. Poor comfort for those who feel their greatest pleasure in the investigation of natural things! Far mistaken is anyone who endeavours to confine chemistry, this noble science, within such narrow bounds! ...

GENERAL PROPERTIES OF ORDINARY AIR

7. (1) Fire must burn for a certain time in a given quantity of air. (2) If, so far as can be seen, this fire does not produce during combustion any fluid resembling air, then, after the fire has gone out of itself, the quantity of air must be diminished between a third and a fourth part. (3) It must not unite with common water. (4) All kinds of animals must live for a certain time in a confined quantity of air. (5) Seeds, as for example peas, in a given quantity of similarly confined air, must strike roots and attain a certain height with the aid of some water and of a moderate heat.

Consequently, when I have a fluid resembling air in its external appearance, and find that it has not the properties mentioned, even when only one of them is wanting, I feel convinced that it is not ordinary air. ...

AIR MUST BE COMPOSED OF ELASTIC FLUIDS OF TWO KINDS

8. *First Experiment.* I dissolved one ounce of alkaline liver of sulphur in eight ounces of water; I poured four ounces of this solution into an empty bottle capable of holding twenty-four ounces of water and closed it most securely with a cork; I then inverted the bottle and placed the neck in a small vessel with water; in this position I allowed it to stand for fourteen days. During this time the solution had lost a part of its red colour and had also deposited some Sulphur: afterwards I took the bottle and held it in the same position in a larger vessel with water, so that the mouth was under and the bottom above the water-level, and withdrew the cork under the water; immediately water rose with violence into the bottle. I closed the bottle again, removed it from the water, and weighed the fluid which it contained. There were ten ounces. After subtracting from this the four ounces of solution of sulphur there remain six ounces, consequently it is apparent from this experiment that of twenty parts of air six parts have been lost in fourteen days. ...

13. *Sixth Experiment.* I collected in a bladder the nitrous air which arises on the dissolution of the metals in acid of nitre, and after I had tied the bladder tightly I laid it in a flask and secured the mouth very carefully with a wet bladder. The nitrous air gradually lost its elasticity, the bladder collapsed, and became yellow as if corroded by *aqua fortis*. After fourteen days I made a hole in the bladder tied over the flask, having previously held it, inverted, under water; the water rose rapidly into the flask, and it remained only two thirds empty. ...

16. It is seen from these experiments that phlogiston, this simple inflammable principle, is present in each of them. It is known that the air strongly attracts to itself the inflammable matter of substances and deprives them of it: not only may this be seen from the experiments cited, but it is at the same time evident that on the transference of the inflammable matter to the air a considerable part of the air is lost. ...

It may also be seen from the above experiments, that a given quantity of air can only unite with, and at the same time saturate, a certain quantity of the inflammable principle: ... Thus much I see from the experiments mentioned, that the air consists of two fluids differing from each other, the one of which does not manifest in the least the property of attracting phlogiston while the other, which composes between the third and the fourth part of the whole mass of the air, is peculiarly disposed to such attraction. But where this latter kind of air has gone to after it has united with the inflammable matter, is a question which must be decided by further experiments, and not by conjectures. ...

First Experiment. I placed nine grains of phosphorus from urine in a thin flask, which was capable of holding thirty ounces of water, and closed its mouth most securely. I then heated, with a burning candle, the part of the flask where the phosphorus lay; the phosphorus began to melt, and immediately afterwards took fire; the flask became filled with a white cloud, which attached itself to the sides like white flowers; this was the dry acid of phosphorus. After the flask had become cold again, I held it, inverted, under water and opened it; scarcely had this been done when the external air pressed water into the flask; this water amounted to nine ounces.

Second Experiment. When I placed pieces of phosphorus in the same flask and allowed it to stand, closed, for six weeks, or until it no longer glowed, I found that one third of the air had been lost.

Third Experiment. I placed three teaspoonfuls of iron filings in a bottle capable of holding two ounces of water; to this I added an ounce of water, and gradually mixed with them half an ounce of oil of vitriol. A violent heating and fermentation took place. When the froth had somewhat subsided, I fixed into the bottle an accurately fitting cork, through which I had previously fixed a glass tube *A* (Fig. 1). I placed this bottle in a vessel filled with hot water, *B B* (cold water would greatly retard the solution). I then approached a burning candle to the orifice of the tube, whereupon the inflammable air took fire and burned with a small yellowish-green flame. As soon as this had taken place, I took a small flask *C* which was capable of holding twenty ounces of water, and held it so deep in the water that the little flame stood in the middle of the flask. The water at once began to rise gradually into the flask, and when the level had reached the point *D* the flame went out. Immediately afterwards the water began to sink again, and was entirely driven out of the flask. The space in the flask up to *D* contained few ounces, therefore the fifth part of the air had been lost. I poured a few ounces of lime water into the flask in order to see whether any aerial acid had also been produced during the combustion, but I did not find any. I made the same experiment with iron filings, and it proceeded in every way similarly to that just mentioned. I shall demonstrate the constituents of this inflammable air further on; for, although it seems to follow from these experiments that it is only phlogiston, still other observations are contrary to this. ...

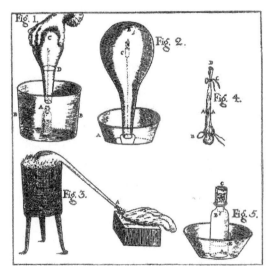

APPARATUS OF SCHEELE.

29. I took a glass retort which was capable of holding eight ounces of water, and distilled fuming acid of nitre according to the usual method. In the beginning the acid went over red, then it became colourless, and finally all became red again; as soon as I perceived the latter, I took away the receiver and tied on a bladder, emptied of air, into which I poured some thick milk of lime in order to prevent the corrosion of the bladder. I then proceeded with the distillation. The bladder began to expand gradually. After this I permitted everything to cool, and tied up the bladder. Lastly I removed it from the neck of the retort. I filled a bottle, which contained ten ounces of water, with this gas. I then placed a small lighted candle in it; scarcely had this been done when the candle began to burn with a large flame, whereby it gave out such a bright light that it was sufficient to dazzle the eyes, I mixed one part of this air with three parts of that kind of air in which fire would not burn; I had here an air which was like the ordinary air in every respect. Since this air is necessarily required for the origination of fire, and makes up about the third part of our common air, I shall call it after this, for the sake of shortness, Fire Air; but the other air which is not in the least serviceable for the fiery phenomenon, and makes up about two thirds of our air, I shall designate after this with the name already known, of Vitiated Air. ...

Electrochemical Researches on the Decomposition of the Earths

With Observations on the Metals Obtained from the Alkaline Earths, and on the Amalgam Procured from Ammonia

by Humphry Davy (1808)

LONDON-BASED HUMPHRY DAVY (1778–1819) spent most of his professional life affiliated with the Royal Society, where he became a public lecturer of great renown. One of the most influential chemists of the early nineteenth century, Davy championed the use of the recently invented and powerful "Voltaic batteries" to isolate new chemical elements. In this excerpt Davy describes the first separation of the highly reactive alkali metals sodium and potassium from their common oxides, potash and soda. His colorful description of the spontaneous explosive burning of these soft silvery elements captures the excitement of these rather dangerous experiments.

* * *

II. ON THE METHODS USED FOR THE DECOMPOSITION OF THE FIXED ALKALIES

The researches I had made on the decomposition of acids, and of alkaline and earthy neutral compounds, proved that the powers of electrical decomposition were proportional to the strength of the opposite electricities in the circuit, and to the conducting power and degree of concentration of the materials employed.

In the first attempts, that I made on the decomposition of the fixed alkalies, I acted upon, aqueous solutions of potash and soda, saturated at common temperatures, by the highest electrical power I could command, and which was produced by a combination of VOLTAIC batteries belonging to the Royal Institution, containing 24 plates of copper and zinc of 12 inches square, 100 plates of 6 inches, and 150 of 4 inches square, charged with solutions of alum and nitrous acid; but in these cases, though there was a high intensity of action, the water of the solutions alone was affected, and hydrogene and oxygene disengaged with the production of much heat and violent effervescence.

The presence of water appearing thus to prevent any decomposition, I used potash in igneous fusion. By means of a stream of oxygene gas from a gasometer applied to the flame of a spirit lamp, which was thrown on a platina spoon containing potash, this alkali was kept for some minutes in a strong red heat, and in a state of perfect fluidity. The spoon was preserved in communication with the positive side of the battery of the power of 100 of 6 inches, highly charged; and the connection from the negative side was made by a platina wire.

By this arrangement some brilliant phenomena were produced. The potash appeared a conductor in a high degree, and as long as the communication was preserved, a most intense light was exhibited at the negative wire, and a column of flame, which seemed to be owing to the development of combustible matter, arose from the point of contact.

When the order was changed, so that the platina spoon was made negative, a vivid and constant light appeared at the opposite point: there was no effect of inflammation round it; but aeriform globules, which inflamed in the atmosphere, rose through the potash.

The platina, as might have been expected, was considerably acted upon; and in the cases when it had been negative, in the highest degree.

The alkali was apparently dry in this experiment; and it seemed probable that the inflammable matter arose from its decomposition. The residual potash was unaltered; it contained indeed a number of dark grey metallic particles, but these proved to be derived from the platina.

I tried several experiments on the electrization of potash rendered fluid by heat, with the hopes of being able to collect the combustible matter, but without success; and, I only attained my object, by employing electricity as the common agent for fusion and decomposition.

Though potash, perfectly dried by ignition, is a non-conductor, yet it is rendered a conductor, by a very slight addition of moisture, which does not perceptibly destroy its aggregation; and in this state it readily fuses and decomposes by strong electrical powers.

A small piece of pure potash, which had been exposed for a few seconds to the atmosphere, so as to give conducting power to the surface, was placed upon an insulated disc of platina, connected with the negative side of the battery of the power of 250 of 6 and 4, in a state of intense activity; and a platina wire, communicating with the positive side, was brought in contact with the upper surface of the alkali. The whole apparatus was in the open atmosphere.

Under these circumstances a vivid action was soon observed to take place. The potash began to fuse at both its points of electrization. There was a violent effervescence at the upper surface; at the lower, or negative surface, there was no liberation of elastic fluid; but small globules having a high metallic lustre, and being precisely similar in visible characters to quicksilver, appeared, some of which burnt with explosion and bright flame, as soon as they were formed, and others remained, and were merely tarnished, and finally covered by a white film which formed on their surfaces.

These globules, numerous experiments soon shewed to be the substance I was in search of, and a peculiar inflammable principle the basis of potash. I found that the platina was in no way connected with the result, except as the medium for exhibiting the electrical powers of decomposition; and a substance of the same kind was produced when pieces of copper, silver, gold, plumbago, or even charcoal were employed for compleating the circuit.

The phenomenon was independent of the presence of air; I found that it took place when the alkali was in the vacuum of an exhausted receiver.

The substance was likewise produced from potash fused by means of a lamp, in glass tubes confined by mercury, and furnished with hermetically inserted platina wires by which the electrical action was transmitted. But this operation could not be carried on for any considerable time; the glass was rapidly dissolved by the action of the alkali, and this substance soon penetrated through the body of the tube.

Soda, when acted upon in the same manner as potash, exhibited an analogous result; but the decomposition demanded greater intensity of action in the batteries, or the alkali was required to be in much thinner and smaller pieces. With the battery of 100 of 6 inches in full activity I obtained good results from pieces of potash weighing from 40 to 70 grains, and of a thickness which made the distance of the electrified metallic surfaces nearly a quarter of an inch; but with a similar power it was impossible to produce the effects of decomposition on pieces of soda of more than 15 or 20 grains in weight, and that only when the distance between the wires was about 1/8 or 1/10 of an inch.

The substance produced from potash remained fluid at the temperature of the atmosphere at the time of its production; that from soda, which was fluid in the degree of heat of the alkali during its formation, became solid on cooling, and appeared having the lustre of silver.

When the power of 250 was used, with a very high charge for the decomposition of soda, the globules often burnt at the moment of their formation, and sometimes violently exploded and separated into smaller globules, which flew with great velocity through the air in a state of vivid combustion, producing a beautiful effect of continued jets of fire.

CHAPTER 11
PROPERTIES OF MATERIALS

*A material's properties result from its constituent atoms and
the arrangements of chemical bonds that hold those atoms together.*

INTRODUCTION

SUPERCONDUCTORS, WHICH CONDUCT electricity without any resistance, are the quintessential high-tech materials. All superconductors require extreme refrigeration to work, but in the late 1980s scientists in Germany, Japan and the United States discovered a new class of materials that superconduct at much higher temperatures.

The research described in *The Breakthrough* was led by physicist Paul Chu of the University of Houston, and included critical contributions by Chu's colleague, Maw Kuen Wu at the University of Alabama. Author Robert Hazen, a crystallographer at the Carnegie Institution of Washington's Geophysical Laboratory, led the effort to solve the atomic structure of the new superconductor. These excerpts come from his first-hand account of the frenetic race to identify and exploit these remarkable new materials.

The Breakthrough:
The Race for the Superconductor

by Robert M. Hazen (1988)

PROLOGUE—JANUARY 1986

SUPERCONDUCTIVITY. THE WORD is magical, just like the phenomenon. Superconductivity is perpetual motion on an atomic scale, the conduction of electricity without the slightest power loss—perfect conductivity. In the first months of 1986 superconductivity was the Holy Grail for a handful of scientists around the world. They sought to comprehend its fascinating physics and tap its potential applications.

But there was one seemingly insurmountable obstacle. Of the hundreds of superconducting materials found prior to January 1986, none would work without refrigeration to a few degrees above absolute zero temperature. (Absolute zero or zero Kelvin, usually abbreviated 0 K, is equal to minus 273 degrees Celsius [written -273°C] and is the lowest possible temperature. Atomic vibration effectively ceases at 0 K.) Superconductor researchers were constantly struggling to find new materials with easily attainable high critical temperatures, or "high Tc," at which superconductivity commences. But, in spite of seventy-five years of research, the high-temperature record has risen only 19 degrees since the original discovery of 4 K mercury superconductivity by Dutch physicist Heike Kamer-lingh-Onnes in 1911.

From 4 K the record had crept upward every few years, a few degrees at a time. But for more than a decade the highest temperature, observed in exotic alloys of the rare metal niobium, remained at about 23 K. A modest industry had arisen for the production and application of powerful superconducting electromagnets fabricated from these alloys. In recent years, however, many scientists had abandoned research on superconductivity. Wishful thinking about commercially viable high-Tc superconductivity was one thing, but hard data from the real world told another, less optimistic story. Many researchers assumed that further substantial advances were unlikely, perhaps even physically impossible.

But Georg Bednorz and Alex Müller, physicists at IBM's Zurich Research Laboratory, didn't give up quite so easily. Rather than study the traditional types of superconductors such as metals and alloys and the like, they followed a hunch and concentrated on oxides—chemical compounds in which the metal atoms are bonded to oxygen. To many it seemed an odd choice. Most oxides are insulators, with no appreciable electrical conductivity, though a couple of odd metal-like oxides had been shown to be superconducting. In 1973, for instance, David Johnson at the University of California, San Diego, observed superconductivity in a lithium-titanium oxide, and two years later Art Sleight at DuPont saw the same effect in an oxide of barium, bismuth, and lead. But those materials had to be cooled to a frigid 13 K, well below the 23 K of recording-holding niobium alloy materials. Few scientists expected higher temperatures from ordinary oxides.

But the two Zurich physicists reasoned differently; they guessed that superconductivity might occur in a small group of odd nickel- or copper-oxygen compounds that displayed peculiar electronic behavior at room temperature. Working without laboratory assistants and with little encouragement from their peers or employers, they mixed elements together in different proportions, cooked the mixtures in high-temperature ovens, and chilled the synthetic samples that emerged, looking in vain for the elusive effect. They labored for two and a half years, synthesizing sample after sample.

Elder partner Karl Alex Müller, with simultaneous senior appointments as professor of physics at the University of Zurich and fellow at IBM's Zurich lab, could work on just about anything he wanted. Sixty years old, with almost a quarter century of IBM service to his credit, Müller had paid his dues. But his younger recruit, thirty-seven-year-old West German

Johann Georg Bednorz, had less freedom. Hired by IBM in 1982, he was paid a good salary to do company work on company projects, not to go off on some half-baked treasure hunt. By agreeing to collaborate with Müller, Bednorz was forced into the uncomfortable situation of secretly synthesizing potential superconductors while simultaneously carrying out his authorized lab duties. As sample after sample failed, the Zurich team's frustration grew.

Every scientific discovery, no matter how revolutionary, builds on earlier findings. The IBM Zurich breakthrough was no exception. In late 1985 Bednorz read a series of papers on unusual copper oxides that had been synthesized by Claude Michel and his coworkers at the University of Caen. In those papers the French workers noted some unusual metal-like electrical conductivity, a property not usually associated with oxides. The Caen group never tested for superconductivity. Bednorz did, and the material worked.

On January 27, 1986, a new era of superconducting science and technology began. Bednorz and Müller, using a variant of the materials synthesized by Michel, smashed the long-standing 23-Kelvin temperature record with a compound of barium, lanthanum, copper, and oxygen that at 30 K displayed the dramatic electrical resistance drop that is a key indicator of superconductivity. The younger Bednorz, frustrated by years of secrecy, was ready to tell the world immediately, but conservative Swiss-born Müller was more cautious. Their result would seem unbelievable to many and, in a field where many claims of high T_c had proven premature, Müller wasn't about to telephone the newspapers. Their scientific reputations were on the line. So Bednorz and Müller cautiously, painstakingly, repeated the experiments until they were absolutely convinced.

Three months passed before Alex Müller dared to show the extraordinary results to anyone. Finally, in mid-April, Müller handed a manuscript with a conservative and unassuming title, "Possible high T_c superconductivity in the Ba-La-Cu-0 system," to his close friend Wolfgang Bückel, who was on the editorial board of *Zeitschrift für Physik*. Though widely read, *Zeit. Phys.* (common shorthand for the monthly) is by no means the most prestigious physics publication.

The journal's advantage to Bednorz and Müller was the confidentiality ensured by their inside contacts on the editorial staff. Bückel, himself a well-known figure in the world of superconductor research, reviewed the paper. And Bednorz's boss, Eric Courtens, also a *Zeit. Phys.* editor, expedited its publication. Even with the paper accepted for publication, Bednorz and Müller did not present the results at any scientific meetings. In a decision that must still give nightmares to IBM executives, they did not even share their history-making data with company colleagues at the large IBM laboratories in New York or California where the findings and their extraordinary ramifications could have been studied in confidence. It took almost half a year more before the historic essay appeared, without fanfare, in the September 1986 issue.

Though no one realized it at the time, this publication marked a historic advance. Not only had Bednorz and Müller eclipsed the previous critical temperature record, but they had done it with a class of superconducting compounds that was all but unknown to workers in the field. By the late fall of 1986 the superconductor rush was about to begin. ...

By the summer of 1986, after having devoted most of his scientific career to the search for superconductors, Paul Chu was becoming less optimistic. Time—and research money—was running out. He had privately resolved, in those hot, humid months of 1986, to abandon superconductors altogether if he couldn't achieve 30 K by 1989, the year of his next grant renewal. Now—suddenly—all those uncertainties were over; Bednorz and Müller had done it. Chu was electrified; he would set his entire laboratory to work at once on the project. He read and reread the paper and then walked across campus to the university library where he nervously passed the time searching for as much related information as he could find until the hour for his research team to show up for work. ...

There are four basic steps in the discovery of a new superconductor. First, obtain a sample, usually through a synthesis process reminiscent of creative cooking: mixing chemicals in careful ratios, grinding them up, and baking until done. In addition to the exact chemical mix, the researcher must determine the optimum temperature of synthesis, the correct gas atmosphere during synthesis, the annealing time at high temperature, and the cooling rate. Sometimes the mix-and-grind technique doesn't work and other complex synthesis procedures involving high-temperature liquids or gases are required. Experienced researchers know that while a single synthesis experiment may take only a few hours, the successful creation of a superconducting sample often requires weeks of patience and creativity.

The second step in establishing the existence of a superconductor is to determine the T_c, the "critical temperature" at which electrical resistance drops to zero. To accomplish this, thin electrical wires are attached to the sample, which is then lowered into an insulated bottle of liquefied gas refrigerant, usually liquid helium. Hoping to see the precipitous drop in resistance, the scientist charts electrical conductivity versus temperature. But a resistance drop by itself can be misleading. A faulty connection or a short circuit can trick the physicist into believing an experiment has been successful. Confidence in the resistance measurements is gained by repeating the experiment while the superconducting sample is subjected to a magnetic field. The T_c *always* drops to a lower temperature in a magnetic field, and scientists must verify that effect in their quest to confirm superconductivity.

The third superconductor test is the demonstration of the "Meissner effect," a subtle phenomenon related to a unique magnetic property of superconductors—the ability to "exclude" a magnetic field. Magnetic fields are all around us, generated by the earth, by permanent magnets, and by electrical devices. These invisible fields pass right through ordinary objects; that's why magnetic fields generated deep within the earth can guide ships and planes on the surface. But superconductors aren't ordinary objects. Magnetic field lines are pushed aside by superconductors. One startling result is that a permanent magnet will float, as if by magic, above a chunk of superconducting material that has been chilled to its critical temperature. ...

The fourth step in verifying that a material is a superconductor is isolation and characterization of the specific superconducting compound. Many superconductor samples consist of a mixture of more than one type of compound, or "phase." Each phase is characterized by its own distinctive chemical composition, atomic structure, and physical properties. Common table salt is a single phase containing sodium and chlorine. A glass of ice water has two phases of H_2O, one solid and one liquid. Most rocks are a cemented mass of different phases (called minerals). A superconducting sample may also have several different phases cemented together, but usually only one of these phases is the superconductor. The physicist's job is to isolate and purify the correct superconducting material and then determine its chemical composition and atomic structure. With that information in hand, engineers can search for applications for the pure superconducting material. ...

The Houston and Alabama researchers continued methodically to produce potential superconducting compounds. Time and again they glimpsed oddities in the electrical resistance suggestive of superconductivity. Transient effects near 100 K were becoming commonplace. Gradually, as evidence became stronger, the Alabama team narrowed its search to a promising composition in the yttrium-barium-copper oxide system—approximately $Y_{1.2}Ba_{0.8}CuO_4$. That composition, the team predicted, would have the highest T_c in the 2-1-4 system. So confident were Wu and his colleagues of success that they abandoned the tedious and expensive liquid helium refrigerant needed to achieve extreme cold and made their preliminary measurements in an open bottle of cheap, 77-K liquid nitrogen.

At 3:00 P.M. on the afternoon of January 29, Maw-Kuen Wu called Paul Chu with an update. The Alabama team had prepared what they hoped would be the best sample thus far. Like the original La-Ba-Cu superconductor, the new Y-Ba-Cu mixture was fine-grained and black, but unlike the other 2-1-4 samples, the Alabama specimen possessed a distinctive greenish cast. A faculty meeting forced Wu to postpone the resistance measurement until 5:00, but Ashburn and Torng made sure everything was ready.

At 5:00 P.M. the measurement was made. Resistivity plummeted at 93 Kelvin. It was no transient effect, and Chu was notified immediately. All of the Alabama and Houston workers were exultant. It was the discovery of a lifetime.

Early on the morning of January 30—the Chinese New Year—Wu flew with the historic sample to Houston for a repeat demonstration of the effect. Given the all too familiar unpredictability of many of their high-T_c samples, there was no guarantee that the results of the previous day could be reproduced. Nerves were on edge as the tiny green-black cut prism was lowered once again into the liquid nitrogen. Once again the yttrium compound did not disappoint. The superconductivity was real.

They all knew the discovery could change the world. It was not just another perfect conductor of electricity. This superconductor worked above 77 K, the temperature of cheap, easy-to-handle liquid nitrogen: 77 K is like the sound barrier or the four-minute mile. It is *the* technological and psychological barrier against which all things cold are gauged. Below

77 K any phenomenon, no matter how remarkable, is an esoteric curiosity with few practical uses. But anyone can buy liquid nitrogen. Above 77 K there are almost no limitations to a material's applications, because almost anything can be refrigerated easily to that temperature. Paul Chu's team had discovered a material that broke the barrier. It could transform superconductivity from an oddity to a day-to-day reality.

CHAPTER 12
THE NUCLEUS OF THE ATOM

Nuclear energy depends on the conversion of mass to energy.

INTRODUCTION

J OSEPH JOHN (J. J.) Thomson (1856–1940) was born in England and, after receiving a degree in engineering, went to Cambridge for advanced training and, later, a faculty appointment. In the series of experiments described in this paper, he reports on an extensive series of experiments on what were then called "cathode rays"—glowing beams that appeared near a negatively charged plate in a vacuum tube. This was at a time when the atom was thought to be something like a bowling ball—completely indivisible—and you can see Thomson struggling to see how to fit his data into a model of the atom. In the end, his experiments established the existence of what we now call the electron and led, eventually, to our modern picture of the atom.

Cathode Rays
by Joseph John Thomson (1897)

THE EXPERIMENTS DISCUSSED in this paper were undertaken in the hope of gaining some information as to the nature of the Cathode Rays. The most diverse opinions are held as to these rays; according to the almost unanimous opinion of German physicists they are due to some process in the aether to which—inasmuch as in a uniform magnetic field their course is circular and not rectilinear—no phenomenon hitherto observed is analogous: another view of these rays is that, so far from being wholly aetherial, they are in fact wholly material, and that they mark the paths of particles of matter charged with negative electricity. It would seem at first sight that it ought not to be difficult to discriminate between views so different, yet experience shows that this is not the case, as amongst the physicists who have most deeply studied the subject can be found supporters of either theory.

The electrified-particle theory has for purposes of research a great advantage over the aetherial theory, since it is definite and its consequences can be predicted; with the aetherial theory it is impossible to predict what will happen under any given circumstances, as on this theory we are dealing with hitherto unobserved phenomena in the aether, of whose laws we are ignorant.

The following experiments were made to test some of the consequences of the electrified-particle theory.

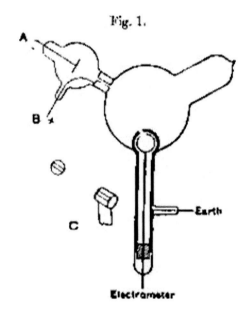

Fig. 1.

CHARGE CARRIED BY THE CATHODE RAYS

If these rays are negatively electrified particles, then when they enter an enclosure they ought to carry into it a charge of negative electricity. This has been proved to be the case by Perrin, who placed in front of a plane cathode two coaxial metallic cylinders which were insulated from each other: the outer of these cylinders was connected with the earth, the inner with a gold-leaf electroscope. These cylinders were closed except for two small holes, one in each cylinder, placed so that the cathode rays could pass through them into the inside of the inner cylinder. Perrin found that when the rays passed into the inner cylinder the electroscope received a charge of negative electricity, while no charge went to the electroscope when the rays were deflected by a magnet so as no longer to pass through the hole.

This experiment proves that something charged with negative electricity is shot off from the cathode, travelling at right angles to it, and that this something is deflected by a magnet; it is open, however, to the objection that it does not prove that the cause of the electrification in the electroscope has anything to do with the cathode rays. Now the supporters of the aetherial theory do not deny that electrified particles are shot off from the cathode; they deny, however, that these charged particles have any more to do with the cathode rays than a rifle-ball has with the flash when a rifle is fired. I have therefore repeated Perrin's experiment in a form which is not open to this objection. The arrangement used was as follows: Two coaxial cylinders (fig. 1) with slits in them are placed in a bulb connected with the discharge-tube; the cathode rays from the cathode A pass into the bulb through a slit in a metal plug fitted into the neck of the tube; this plug is connected with the anode and is put to earth. The cathode rays thus do not fall upon the cylinders unless they are deflected by a magnet. The outer cylinder is connected with the earth, the inner with the electrometer. When the cathode rays (whose path was traced by the phosphorescence on the glass) did not fall on the slit, the electrical charge sent to the electrometer when the induction-coil producing the rays was set in action was small and irregular; when, however, the rays were bent by a magnet so as to fall on the slit there was a large charge of negative electricity sent to the electrometer. I was surprised at the magnitude of the charge; on some occasions enough negative electricity went through the narrow slit into the inner cylinder in one second to alter the potential of a capacity of 1.5 microfarads by 20 volts. If the rays were so much bent by the magnet that they overshot the slits in the cylinder, the charge passing into the cylinder fell again to a very small fraction of its value when the aim was true. Thus this experiment shows that however we twist and deflect the cathode rays by magnetic forces, the negative electrification follows the same path as the rays, and that this negative electrification is indissolubly connected with the cathode rays.

DEFLEXION OF THE CATHODE RAYS BY AN ELECTROSTATIC FIELD

An objection very generally urged against the view that the cathode rays are negatively electrified particles, is that hitherto no deflexion of the rays has been observed under a small electrostatic force, and though the rays are deflected when they pass near electrodes connected with sources of large differences of potential, such as induction-coils or electrical machines, the deflexion in this case is regarded by the supporters of the aetherial theory as due to the discharge passing between the electrodes, and not primarily to the electrostatic field. Hertz made the rays travel between two parallel plates of metal placed inside the discharge-tube, but found that they were not deflected when the plates were connected with a battery of storage-cells; on repeating this experiment I at first got the same result, but subsequent experiments showed that the absence of deflexion is due to the conductivity conferred on the rarefied gas by the cathode rays. On measuring this conductivity it was found that it diminished very rapidly as the exhaustion increased; it seemed then that on trying Hertz's experiment at very high exhaustions there might be a chance of detecting the deflexion of the cathode rays by an electrostatic force.

The apparatus used is represented in fig. 2.

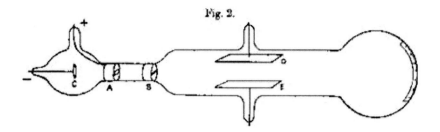

Fig. 2.

The rays from the cathode C pass through a slit in the anode A, which is a metal plug fitting tightly into the tube and connected with the earth; after passing through a second slit in another earth-connected metal plug B, they travel between two parallel aluminium plates about 5 cm. long by 2 broad and at a distance of 1.5 cm. apart; they then fall on the end of the tube and produce a narrow well-defined phosphorescent patch. A scale pasted on the outside of the tube serves to measure the deflexion of this patch. At high exhaustions the rays were deflected when the two aluminium plates were connected with the terminals of a battery of small storage cells; the rays were depressed when the upper plate was connected with the negative pole of the battery, the lower with the positive, and raised when the upper plate was connected with the positive, the lower with the negative pole. The deflexion was proportional to the difference of potential between the plates, and I could detect the deflexion when the potential-difference was as small as two volts.

[*EDITOR'S NOTE: This is followed by a long discussion of the deflection of cathode rays (we would call them electrons) by magnetic fields and their behavior in different kinds of gasses.*]

The two fundamental points about these carriers seem to me to be (1) that these carriers are the same whatever the gas through which the discharge passes, (2) that the mean free paths depend upon nothing but the density of the medium traversed by these rays.

The explanation which seems to me to account in the most simple and straightforward manner for the facts is founded on a view of the constitution of the chemical elements which has been favourably entertained by many chemists: this view is that the atoms of the different chemical elements are different aggregations of atoms of the same kind. In the form in which this hypothesis was enunciated by Prout, the atoms of the different elements were hydrogen atoms; in this precise form the hypothesis is not tenable, but if we substitute for hydrogen some unknown primordial substance X, there is nothing known which is inconsistent with this hypothesis, which is one that has been recently supported by Sir Norman Lockyer for reasons derived from the study of the stellar spectra.

If, in the very intense electric field in the neighbourhood of the cathode, the molecules of the gas are dissociated and are split up, not into the ordinary chemical atoms, but into these primordial atoms, which we shall for brevity call corpuscles; and if these corpuscles are charged with electricity and projected from the cathode by the electric field, they would behave exactly like the cathode rays. They would evidently give a value of m/e which is independent of the nature of the gas and its pressure, for the carriers are the same whatever the gas may be; again, the mean free paths of these corpuscles would depend solely upon the density of the medium through which they pass. For the molecules of the medium are composed of a number of such corpuscles separated by considerable spaces; now the collision between a single corpuscle and the molecule will not be between the corpuscles and the molecule as a whole, but between this corpuscle and the individual corpuscles which form the molecule; thus the number of collisions the particle makes as it moves through a crowd of these molecules will be proportional, not to the number of the molecules in the crowd, but to the number of the individual corpuscles. The mean free path is inversely proportional to the number of collisions in unit time, and so is inversely proportional to the number of corpuscles in unit volume; now as these corpuscles are all of the same mass, the number of corpuscles in unit volume will be proportional to the mass of unit volume, that is the mean free path will be inversely proportional to the density of the gas. We see, too, that so long as the distance between neighbouring corpuscles is large compared with the linear dimensions of a corpuscle the mean free path will be independent of the way they are arranged, provided the number in unit volume remains constant, that is the mean free path will depend only on the density of the medium traversed by the corpuscles, and will be independent of its chemical nature and physical state: this from Lenard's very remarkable measurements of the absorption of the cathode rays by various media, must be a property possessed by the carriers of the charges in the cathode rays.

Thus on this view we have in the cathode rays matter in a new state, a state in which the subdivision of matter is carried very much further than in the ordinary gaseous state: a state in which all matter—that is, matter derived from different sources such as hydrogen, oxygen, &c.—is of one and the same kind; this matter being the substance from which all the chemical elements are built up.

VELOCITY OF THE CATHODE RAYS

The velocity of the cathode rays is variable, depending upon the potential-difference between the cathode and anode, which is a function of the pressure of the gas—the velocity increases as the exhaustion improves; the measurements given above show, however, that at all the pressures at which experiments were made the velocity exceeded 10^9 cm./sec. This velocity is much greater than the value of 2×10^7 which I previously obtained (Phil. Mag. Oct. 1894) by measuring directly the interval which separated the appearance of luminosity at two places on the walls of the tube situated at different distances from the cathode.

I have much pleasure in thanking Mr. Everitt for the assistance he has given me in the preceding investigation.

The Scattering of α and β Particles by Matter and the Structure of the Atom

by Ernest Rutherford (1911)

ERNEST RUTHERFORD (1871–1937) was born in New Zealand and educated at Canterbury College, Christ Church, and Cambridge, England. Moving to a professorship at McGill University in Montreal, he investigated the nature of the new phenomenon of radioactivity, and identified the alpha particle as the nucleus of the helium atom—work for which he was awarded the Nobel Prize in chemistry in 1908. (He is probably the only person in history to do his most important work *after* receiving this award.) After moving to the University of Manchester in England, he did the work reported in this paper, which established the existence of the atomic nucleus and therefore gave us our modern picture of the atom. The material we have edited out were mainly lengthy calculations showing that the scattering he observed was due to a single collision of an alpha particle with an atom, and not a series of small angle collisions.

* * *

§ 1.

It is well known that the α and the β particles suffer deflexions from their rectilinear paths by encounters with atoms of matter. This scattering is far more marked for the β than for the α particle on account of the much smaller momentum and energy of the former particle. There seems to be no doubt that such swiftly moving particles pass through the atoms in their path, and that the deflexions observed are due to the strong electric field traversed within the atomic system. It has generally been supposed that the scattering of a pencil of α or β rays in passing through a thin plate of matter is the result of a multitude of small scatterings by the atoms of matter traversed. The observations, however, of Geiger and Marsden[2] on the scattering of α rays indicate that some of the α particles, about 1 in 20,000 were turned through an average angle of 90 degrees in passing though a layer of gold-foil about 0.00004 cm. thick, which was equivalent in stopping-power of the α particle to 1.6 millimetres of air. Geiger[3] showed later that the most probable angle of deflexion for a pencil of α particles being deflected through 90 degrees is vanishingly small. In addition, it will be seen later that the distribution of the α particles for various angles of large deflexion does not follow the probability law to be expected if such large deflexion are made up of a large number of small deviations. It seems reasonable to suppose that the deflexion through a large angle is due to a single atomic encounter, for the chance of a second encounter of a kind to produce a large deflexion must in most cases be exceedingly small. A simple calculation shows that the atom must be a seat of an intense electric field in order to produce such a large deflexion at a single encounter.

Recently Sir J. J. Thomson has put forward a theory to explain the scattering of electrified particles in passing through small thicknesses of matter. The atom is supposed to consist of a number N of negatively charged corpuscles, accompanied by an equal quantity of positive electricity uniformly distributed throughout a sphere. The deflexion of a negatively electrified particle in passing through the atom is ascribed to two causes—(1) the repulsion of the corpuscles distributed through the atom, and (2) the attraction of the positive electricity in the atom. The deflexion of the particle in passing through the atom is supposed to be small, while the average deflexion after a large number m of encounters was taken as [the square root of] m θ, where θ is the average deflexion due to a single atom. It was shown that the number N of the electrons within the atom could be deduced from observations of the scattering was examined experimentally by Crowther in a later paper. His results apparently confirmed the main conclusions of the theory, and he deduced, on the assumption that the positive electricity was continuous, that the number of electrons in an atom was about three times its atomic weight.

The theory of Sir J. J. Thomson is based on the assumption that the scattering due to a single atomic encounter is small, and the particular structure assumed for the atom does not admit of a very large deflexion of diameter of the sphere of positive electricity is minute compared with the diameter of the sphere of influence of the atom.

Since the α and β particles traverse the atom, it should be possible from a close study of the nature of the deflexion to form some idea of the constitution of the atom to produce the effects observed. In fact, the scattering of high-speed charged particles by the atoms of matter is one of the most promising methods of attack of this problem. The development of the scintillation method of counting single α particles affords unusual advantages of investigation, and the researches of H. Geiger by this method have already added much to our knowledge of the scattering of α rays by matter.

§ 2.

We shall first examine theoretically the single encounters with an atom of simple structure, which is able to produce large deflections of an α particle, and then compare the deductions from the theory with the experimental data available.

Consider an atom which contains a charge ±Ne at its centre surrounded by a sphere of electrification containing a charge ±Ne supposed uniformly distributed throughout a sphere of radius R. e is the fundamental unit of charge, which in this paper is taken as 4.65 x 10^{-10} E.S. unit. We shall suppose that for distances less than 10^{-12} cm. the central charge and also the charge on the alpha particle may be supposed to be concentrated at a point. It will be shown that the main deductions from the theory are independent of whether the central charge is supposed to be positive or negative. For convenience, the sign will be assumed to be positive. The question of the stability of the atom proposed need not be considered at this stage, for this will obviously depend upon the minute structure of the atom, and on the motion of the constituent charged parts.

In order to form some idea of the forces required to deflect an alpha particle through a large angle, consider an atom containing a positive charge Ne at its centre, and surrounded by a distribution of negative electricity Ne uniformly distributed within a sphere of radius R. The electric force X and the potential V at a distance r from the centre of an atom for a point inside the atom, are given by

$$X = Ne\left(\frac{1}{r^2} - \frac{r}{R^3}\right)$$

$$V = Ne\left(\frac{1}{r} - \frac{3}{2R} + \frac{r^2}{2R^3}\right).$$

Suppose an α particle of mass m and velocity u and charge E shot directly towards the centre of the atom. It will be brought to rest at a distance b from the centre given by

$$\tfrac{1}{2}mu^2 = NeE\left(\frac{1}{b} - \frac{3}{2R} + \frac{b^2}{2R^3}\right).$$

It will be seen that b is an important quantity in later calculations. Assuming that the central charge is 100 e, it can be calculated that the value of b for an α particle of velocity 2.09×10^9 cms. per second is about 3.4×10^{-12} cm. In this calculation b is supposed to be very small compared with R. Since R is supposed to be of the order of the radius of the atom, viz. 10^{-8} cm., it is obvious that the α particle before being turned back penetrates so close to the central charge, that the field due to the uniform distribution of negative electricity may be neglected. In general, a simple calculation shows that for all deflexions greater than a degree, we may without sensible error suppose the deflexion due to the field of the central charge alone. Possible single deviations due to the negative electricity, if distributed in the form of corpuscles, are not taken into account at this stage of the theory. It will be shown later that its effect is in general small compared with that due to the central field.

Consider the passage of a positive electrified particle close to the centre of an atom. Supposing that the velocity of the particle is not appreciably changed by its passage through the atom, the path of the particle under the influence of a repulsive force varying inversely as the square of the distance will be an hyperbola with the centre of the atom S as the external focus. Suppose the particle to enter the atom in the direction PO (fig. 1), and that the direction of motion on escaping the atom is OP'. OP and OP' make equal angles with the line SA, where A is the apse of the hyperbola. p = SN = perpendicular distance from centre on direction of initial motion of particle.

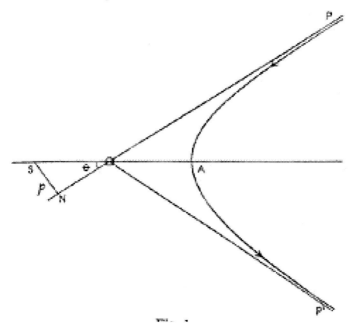

[EDITOR'S NOTE: *At this point, Rutherford works out the trajectory and angle of deflection of the particle scattering from the atom. He then works out the probability of a large angle scattering being the result of a single collision. He also works out how much the alpha particles will be slowed down by the collision.*]

In these calculations, it is assumed that the α particles scattered through a large angle suffer only one large deflexion. For this to hold, it is essential that the thickness of the scattering material should be so small that the chance of a second encounter involving another large deflexion is very small. If, for example, the probability of a single deflexion φ in passing through a thickness t is 1/1000, the probability of two successive deflexions each of value φ is $1/10^6$, and is negligibly small.

The angular distribution of the α particles scattered from a thin metal sheet affords one of the simplest methods of testing the general correctness of this theory of single scattering. This has been done recently for α rays by Dr. Geiger, who found that the distribution for particles deflected between 30° and 150° from a thin gold-foil was in substantial agreement with the theory. A more detailed account of these and other experiments to test the validity of the theory will be published later.

It is seen that the reduction of velocity of the α particle becomes marked on this theory for encounters with the lighter atoms. Since the range of an α particle in air or other matter is approximately proportional to the cube of the velocity, it follows that an α particle of range 7 cms. has its range reduced to 4.5 cms. after incurring a single deviation of 90° in traversing an aluminium atom. This is of a magnitude to be easily detected experimentally. Since the value of K is very large for an encounter of a β particle with an atom, the reduction of velocity on this formula is very small.

Some very interesting cases of the theory arise in considering the changes of velocity and the distribution of scattered particles when the α particle encounters a light atom, for example a hydrogen or helium atom. A discussion of these and similar cases is reserved until the question has been examined experimentally.

[EDITOR'S NOTE: *Rutherford then gives mathematical arguments to show that alpha particles are scattered through large angles by a single collision, rather than by multiple small collisions.*]

It is evident from this comparison, that the probability for any given deflexion is always greater for single than for compound scattering. The difference is especially marked when only a small fraction of the particles are scattered through any given angle. It follows from this result that the distribution of particles due to encounters with the atoms is for small thicknesses mainly governed by single scattering. No doubt compound scattering produces some effect in equalizing the distribution of the scattered particles; but its effect becomes relatively smaller, the smaller the fraction of the particles scattered through a given angle.

§6. COMPARISON OF THEORY WITH EXPERIMENTS

On the present theory, the value of the central charge Ne is an important constant, and it is desirable to determine its value for different atoms. This can be most simply done by determining the small fraction of α or β particles of known velocity falling on a thin metal screen, which are scattered between φ and φ + dφ where φ is the angle of deflexion, The influence of compound scattering should be small when this fraction is small.

Experiments in these directions are in progress, but it is desirable at this stage to discuss in the light of the present theory the data already published on scattering of α and β particles,

The following points will be discussed: —

(a) The "diffuse reflexion" of α particles, i.e. the scattering of α particles through large angles (Geiger and Marsden.)

(b) The variation of diffuse reflexion with atomic weight of the radiator (Geiger and Marsden.)

(c) The average scattering of a pencil of α rays transmitted through a thin metal plate (Geiger.)

(a) In the paper of Geiger and Marsden (loc.cit.) on the diffuse reflexion of α particles falling on various substances it was shown that about 1/8000 of the α particles from radium C falling on a thick plate of platinum are scattered back in the direction of the incidence. This fraction is deduced on the assumption that the α particles are uniformly scattered in all directions , the observation being made for a deflexion of about 90°. The form of experiment is not very suited for accurate calculation, but from the data available it can be shown that the scattering observed is about that to be expected on the theory if the atom of platinum has a central charge of about 100 e.

In their experiments on this subject, Geiger and Marsden gave the relative number of α particles diffusely reflected from thick layers of different metals, under similar conditions. The numbers obtained by them are given in the table below, where z represents the relative number of scattered particles, measured by the number of scintillations per minute on a zinc sulphide screen.

Metal	Atomic Weight	z	$z/A^{3/2}$
Lead	207	62	208
Gold	197	67	242
Platinum	195	63	232
Tin	119	34	226
Silver	108	27	241
Copper	64	14.5	225
Iron	56	10.2	250
Aluminium	27	3.4	243
			Average 233

On the theory of single scattering, the fraction of the total number of α particles scattered through any given angle in passing through a thickness t is proportional to nA^2t , assuming that the central charge is proportional to the atomic weight A. In the present case, the thickness of matter from which the scattered α particles are able to emerge and affect the zinc sulphide screen depends on the metal. Since Bragg has shown that the stopping power of an atom for an α particle is proportional to the square root of its atomic weight, the value of nt for different elements is proportional to 1 / [square root of] A . In this case t represents the greatest depth from which the scattered α particles emerge. The number z of α particles scattered back from a thick layer is consequently proportional to $A^{3/2}$ or $z / A^{3/2}$ should be a constant.

To compare this deduction with experiment, the relative values of the latter quotient are given in the last column. Considering the difficulty of the experiments, the agreement between theory and experiment is reasonably good.

The single large scattering of α particles will obviously affect to some extent the shape of the Bragg ionization curve for a pencil of α rays. This effect of large scattering should be marked when the α rays have traversed screens of metals of high atomic weight, but should be small for atoms of light atomic weight.

(c) Geiger made a careful determination of the scattering of α particles passing through thin metal foils, by the scintillation method, and deduced the most probable angle through which the α particles are deflected in passing through known thickness of different kinds of matter.

A narrow pencil of homogeneous α rays was used as a source. After passing through the scattering foil, the total number of α particles are deflected through different angles was directly measured. The angle for which the number of scattered particles was a maximum was taken as the most probable angle. The variation of the most probable angle with thickness of matter was determined, but calculation from these data is somewhat complicated by the variation of velocity of the α particles in their passage through the scattering material. A consideration of the curve of distribution of the α particles given in the paper (loc.cit. p. 498) shows that the angle through which half the particles are scattered is about 20 per cent greater than the most probable angle. …

Geiger showed that the most probable angle of deflexion for an atom was nearly proportional to its atomic weight. It consequently follows that the value for N for different atoms should be nearly proportional to their atomic weights, at any rate for atomic weights between gold and aluminum.

Since the atomic weight of platinum is nearly equal to that of gold, it follows from these considerations that the magnitude of the diffuse reflexion of α particles through more than 90° from gold and the magnitude of the average small angle scattering of a pencil of rays in passing through gold-foil are both explained on the hypothesis of single scattering by supposing the atom of gold has a central charge of about 100 e. …

We have already seen that the distribution of scattered α particles at various angles has been found by Geiger to be in substantial agreement with the theory of single scattering, but cannot be explained on the theory of compound scattering alone. Since there is every reason to believe that the laws of scattering of α and β particles are very similar, the law of distribution of scattered β particles should be the same as for α particles for small thicknesses of matter. …

§7. GENERAL CONSIDERATIONS

In comparing the theory outlined in this paper with the experimental results, it has been supposed that the atom consists of a central charge supposed concentrated at a point, and that the large single deflexions of the α and β particles are mainly due to their passage through the strong central field. The effect of the equal and opposite compensation charge supposed distributed uniformly throughout a sphere has been neglected. Some of the evidence in support of these assumptions will now be briefly considered. For concreteness, consider the passage of a high speed α particle through an atom having a positive central charge Ne, and surrounded by a compensating charge of N electrons. Remembering that the mass, momentum, and kinetic energy of the α particle are very large compared with the corresponding values of an electron in rapid motion, it does not seem possible from dynamic considerations that an α particle can be deflected through a large angle by a close approach to an electron, even if the latter be in rapid motion and constrained by strong electrical forces. It seems reasonable to suppose that the chance of single deflexions through a large angle due to this cause, if not zero, must be exceedingly small compared with that due to the central charge.

It is of interest to examine how far the experimental evidence throws light on the question of extent of the distribution of central charge. Suppose, for example, the central charge to be composed of N unit charges distributed over such a volume that the large single deflexions are mainly due to the constituent charges and not to the external field produced by the distribution. It has been shown (§3) that the fraction of the α particles scattered through a large angle is proportional to $(NeE)^2$, where Ne is the central charge concentrated at a point and E the charge on the deflected particles. If, however, this charge is distributed in single units, the fraction of the α particles scattered through a given angle is proportional of Ne^2 instead of N^2e^2. In this calculation, the influence of mass of the constituent particle has been neglected, and account has only been taken of its electric field. Since it has been shown that the value of the central point charge for gold must be about 100, the value of the distributed charge required to produce the same proportion of single deflexions through a large angle should be at least 10,000. Under these conditions the mass of the constituent particle would be small compared with that of the α particle, and the difficulty arises of the production of large single deflexions at all. In addition, with such a large distributed charge, the effect of compound scattering is relatively more important than that of single scattering. For example, the probable small angle of deflexion of pencil of α particles passing through a thin gold foil would be much greater than that experimentally observed by Geiger (§ b–c). The large and small angle scattering could not then be explained by the assumption of a central charge of the same value. Considering the evidence as a whole, it seems simplest to suppose that the atom contains a central charge distributed through a very small volume, and that the large single deflexions are due to the central charge as a whole, and not to its constituents. At the same time, the experimental evidence is not precise enough to negate the possibility that a small fraction of the positive charge may be carried by satellites extending some distance from the centre. Evidence on this point could be obtained by examining whether the same central charge is required to explain the large single deflexions of α and β particles; for the α particle must approach much closer to the center of the atom than the β particle of average speed to suffer the same large deflexion.

The general data available indicate that the value of this central charge for different atoms is approximately proportional to their atomic weights, at any rate of atoms heavier than aluminium. It will be of great interest to examine experimentally whether such a simple relation holds also for the lighter atoms. In cases where the mass of the deflecting atom (for example, hydrogen, helium, lithium) is not very different from that of the α particle, the general theory of single scattering will require modification, for it is necessary to take into account the movements of the atom itself (see § 4). ...

The deductions from the theory so far considered are independent of the sign of the central charge, and it has not so far been found possible to obtain definite evidence to determine whether it be positive or negative. It may be possible to settle the question of sign by consideration of the difference of the laws of absorption of the β particles to be expected on the two hypothesis, for the effect of radiation in reducing the velocity of the β particle should be far more marked with a positive than with a negative center. If the central charge be positive, it is easily seen that a positively charged mass if released from the center of a heavy atom, would acquire a great velocity in moving through the electric field. It may be possible in this way to account for the high velocity of expulsion of α particles without supposing that they are initially in rapid motion within the atom.

Further consideration of the application of this theory to these and other questions will be reserved for a later paper, when the main deductions of the theory have been tested experimentally. Experiments in this direction are already in progress by Geiger and Marsden.

CHAPTER 13
THE ULTIMATE STRUCTURE
OF MATTER

All matter is made of quarks and leptons,
which are the fundamental building blocks of the universe.

INTRODUCTION

T HE QUEST FOR the most fundamental building blocks of nature has been a focus of physicists for centuries. Among the most fascinating discoveries of 20th century science were many new subatomic particles, including quarks (the building blocks of nuclear particles), leptons (including neutrinos), and antimatter.

Carl Anderson (1905–1991) was born in New York City and studied at the California Institute of Technology, where he spent his entire career. In this paper he reports the first evidence for what we now call antimatter. In the early 1930s, the only way that physicists could study the structure of the nucleus was to see what happens when atoms were hit by high energy cosmic rays. These rays consist mostly of protons that have been accelerated to high energy somewhere in the cosmos, and rain down on the Earth continuously. In a typical experiment, a detection device would be placed on a mountaintop (so the cosmic rays didn't have to pass through a lot of the atmosphere). Anderson's device was called a cloud chamber and it worked this way: as particles passed through or collided, they left behind them a string of ionized atoms. These ions served as nuclei for the condensation of droplets, which were then photographed to reveal the particle's track. Note Anderson's (successful) introduction of the term "positron" for his new particle, and his (unsuccessful) attempt to change the name of the electron to "negatron."

The Positive Electron
by Carl D. Anderson (1933)

ON AUGUST 2, 1932, during the course of photographing cosmic-ray tracks produced in a vertical Wilson chamber (magnetic field of 15,000 gauss) designed in the summer of 1930 by Professor R. A. Millikan and the writer, the tracks shown in Fig. 1 were obtained, which seemed to be interpretable only on the basis of the existence in this case of a particle carrying a positive charge but having a mass of the same order of magnitude as that normally possessed by a free negative electron. Later study of the photograph by a whole group of men of the Norman Bridge Laboratory only tended to strengthen this view. The reason that this interpretation seemed so inevitable is that the track appearing on the upper half of the figure cannot possibly have a mass as large as that of a proton for as soon as the mass is fixed the energy is at once fixed by the curvature. The energy of a proton of that curvature comes out 300,000 volts, but a proton of that energy according to well established and universally accepted determinations[1] has a total range of about 5 mm in air while that portion of the range actually visible in this case exceeds 5 cm without a noticeable change in curvature. The only escape from this conclusion would be to assume that at exactly the same instant (and the sharpness of the tracks determines that instant to within about a fiftieth of a second) two independent electrons happened to produce two tracks so placed as to give the impression of a single particle shooting through the lead plate. This assumption was dismissed on a probability basis, since a sharp track of this order of curvature under the experimental conditions prevailing occurred in the chamber only once in some 500 exposures, and since there was practically no chance at all that two such tracks should line up in this way. We also discarded as completely untenable the assumption of an electron of 20 million volts entering the lead on one side and coming out with an energy of 60 million volts on the other side. A fourth possibility is that a photon, entering the lead from above, knocked out of the nucleus of a lead atom two particles, one of which shot upward and the other downward. But in this case the upward moving one would be a positive of small mass so that either of the two possibilities leads to the existence of the positive electron.

In the course of the next few weeks other photographs were obtained which could be interpreted logically only on the positive-electron basis, and a brief report was then published[2] with due reserve in interpretation in view of the importance and striking nature of the announcement.

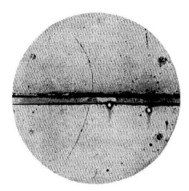

FIG. 1. A 63 million volt positron ($H\rho = 2.1 \times 10^5$ gauss-cm) passing through a 6 mm lead plate and emerging as a 23 million volt positron ($H\rho = 7.5 \times 10^4$ gauss-cm). The length of this latter path is at least ten times greater than the possible length of a proton path of this curvature.

MAGNITUDE OF CHARGE AND MASS

It is possible with the present experimental data only to assign rather wide limits to the magnitude of the charge and mass of the particle. The specific ionization was not in these cases measured, but it appears very probable, from a knowledge of the experimental conditions and by comparison with many other photographs of high- and low-speed electrons taken under the same conditions, that the charge cannot differ in magnitude from that of an electron by an amount as great as a factor of two. Furthermore, if the photograph is taken to represent a positive particle penetrating the 6 mm lead plate, then the energy lost, calculated for unit charge, is approximately 38 million electron-volts, this value being practically independent of the proper mass of the particle as long as it is not too many times larger than that of a free negative electron. This value of 63 million volts per cm energy-loss for the positive particle it was considered legitimate to compare with the measured mean of approximately 35 million volts[3] for negative electrons of 200–300 million volts energy since the rate of energy-loss for particles of small mass is expected to change only very slowly over an energy range extending from several million to several hundred million volts. Allowance being made for experimental uncertainties, an upper limit to the rate of loss of energy for the positive particle can then be set at less than four times that for an electron, thus fixing, by the usual relation between rate of ionization and charge, an upper limit to the charge less than twice that of the negative electron. It is concluded, therefore, that the magnitude of the charge of the positive electron which we shall henceforth contract to positron is very probably equal to that of a free negative electron which from symmetry considerations would naturally then be called a negatron.

To date, out of a group of 1300 photographs of cosmic-ray tracks 15 of these show positive particles penetrating the lead, none of which can be ascribed to particles with a mass as large as that of a proton, thus establishing the existence of positive particles of unit charge and of mass small compared to that of a proton. In many other cases due either to the short section of track available for measurement or to the high energy of the particle it is not possible to differentiate with certainty between protons and positrons. A comparison of the six or seven hundred positive-ray tracks which we have taken is, however, still consistent with the view that the positive particle which is knocked out of the nucleus by the incoming primary cosmic ray is in many cases a proton.

From the fact that positrons occur in groups associated with other tracks it is concluded that they must be secondary particles ejected from an atomic nucleus. If we retain the view that a nucleus consists of protons and neutrons (and *a*-particles) and that a neutron represents a close combination of a proton and electron, then from the electromagnetic theory as to the origin of mass the simplest assumption would seem to be that an encounter between the incoming primary ray and a proton may take place in such a way as to expand the diameter of the proton to the same value as that possessed by the negatron. This process would release an energy of a billion electron-volts appearing as a secondary photon. As a second possibility the primary ray may disintegrate a neutron (or more than one) in the nucleus by the ejection either of a negatron or a positron with the result that a positive or a negative proton, as the case may be, remains in the nucleus in place of the neutron, the event occurring in this instance without the emission of a photon. This alternative, however, postulates the existence in the nucleus of a proton of negative charge, no evidence for which exists. The greater symmetry, however, between the positive and negative charges revealed by the discovery of the positron should prove a stimulus to search for evidence of the existence of negative protons. If the neutron should prove to be a fundamental particle of a new kind rather than a proton and negatron in close combination, the above hypotheses will have to be abandoned for the proton will then in all probability be represented as a complex particle consisting of a neutron and positron. ...

Fig. 2. A positron of 20 million volts energy (Hρ = 7.1 × 10⁴ gauss-cm) and a negatron of 30 million volts energy (Hρ = 10.2 × 10⁴ gauss-cm) projected from a plate of lead. The range of the positive particle precludes the possibility of ascribing it to a proton of the observed curvature.

Fig. 3. A group of six particles projected from a region in the wall of the chamber. The track at the left of the central group of four tracks is a negatron of about 18 million volts energy (Hρ = 6.2 × 10⁴ gauss-cm) and that at the right a positron of about 20 million volts energy (Hρ = 7.0 × 10⁴ gauss-cm). Identification of the two tracks in the center is not possible. A negatron of about 15 million volts is shown at the left. This group represents early tracks which were broadened by the diffusion of the ions. The uniformity of this broadening for all the tracks shows that the particles entered the chamber at the same time.

Fig. 4. A positron of about 200 million volts energy (Hρ = 6.6 × 10⁵ gauss-cm) penetrates the 11 mm lead plate and emerges with about 125 million volts energy (Hρ = 4.2 × 10⁵ gauss-cm). The assumption that the tracks represent a proton traversing the lead plate is inconsistent with the observed curvatures. The energies would then be, respectively, about 20 million and 8 million volts above and below the lead, energies too low to permit the proton to have a range sufficient to penetrate a plate of lead of 11 mm thickness.

Large Hadron Collider: The Discovery Machine
by Graham P. Collins (2008)

SINCE THE 1930S, progress in our understanding of the basic structure of matter has depended on our ability to build ever more powerful machines called accelerators. These machines accelerate particles such as protons or electrons to extremely high energies, they smash them into other particles or atoms. From the debris of these collisions, scientists try to deduce the nature of the target. For the last several decades of the 20th century, the world's largest accelerator was at the Fermi National Accelerator Laboratory outside of Chicago. At the start of this century, however, all eyes turned toward Geneva, Switzerland, where a new machine called the Large Hadron Collider was being assembled. This article gives some sense of the enormous complexity as well as the enormous hopes for this new instrument.

An update on the progress of the machine: in September, 2008, the first beam was successfully sent around the LHC. Shortly thereafter, however, an electrical connection between two magnets burned out, causing an explosion that damaged the beam pipe and 53 of the machine's magnets. Repairs were completed in 2009, and as of this writing the scientific world is waiting for the next ramping up of the LHC. It is not unusual for complex machines like the LHC to have difficulties at the beginning—the Hubble Space Telescope, for example, needed repairs after launch, but quickly became the most productive astronomical instrument of all time. In the words of rocket engineer Werner von Braun, "If a machine works the first time, it's over-designed."

* * *

You could think of it as the biggest, most powerful microscope in the history of science. The Large Hadron Collider (LHC), now being completed underneath a circle of countryside and villages a short drive from Geneva, will peer into the physics of the shortest distances (down to a nano-nanometer) and the highest energies ever probed. For a decade or more, particle physicists have been eagerly awaiting a chance to explore that domain, sometimes called the terascale because of the energy range involved: a trillion electron volts, or 1 TeV. Significant new physics is expected to occur at these energies, such as the elusive Higgs particle (believed to be responsible for imbuing other particles with mass) and the particle that constitutes the dark matter that makes up most of the material in the universe.

The mammoth machine, after a nine-year construction period, is scheduled (touch wood) to begin producing its beams of particles later this year. The commissioning process is planned to proceed from one beam to two beams to colliding beams; from lower energies to the terascale; from weaker test intensities to stronger ones suitable for producing data at useful rates but more difficult to control. Each step along the way will produce challenges to be overcome by the more than 5,000 scientists, engineers and students collaborating on the gargantuan effort. When I visited the project last fall to get a firsthand look at the preparations to probe the high-energy frontier, I found that everyone I spoke to expressed quiet confidence about their ultimate success, despite the repeatedly delayed schedule. The particle physics community is eagerly awaiting the first results from the LHC. Frank Wilczek of the Massachusetts Institute of Technology echoes a common sentiment when he speaks of the prospects for the LHC to produce "a golden age of physics."

A MACHINE OF SUPERLATIVES

To break into the new territory that is the terascale, the LHC's basic parameters outdo those of previous colliders in almost every respect. It starts by producing proton beams of far higher energies than ever before. Its nearly 7,000 magnets, chilled by liquid helium to less than two kelvins to make them superconducting, will steer and focus two beams of protons traveling within a millionth of a percent of the speed of light. Each proton will have about 7 TeV of energy—7,000 times as much energy as a proton at rest has embodied in its mass, courtesy of Einstein's $E = mc^2$. That is about seven times the energy of the reigning record holder, the Tevatron collider at Fermi National Accelerator Laboratory in Batavia, Ill. Equally important, the machine is designed to produce beams with 40 times the intensity, or luminosity, of the Tevatron's beams. When it is fully loaded and at maximum energy, all the circulating particles will carry energy roughly equal to the kinetic energy of about 900 cars traveling at 100 kilometers per hour, or enough to heat the water for nearly 2,000 liters of coffee.

The protons will travel in nearly 3,000 bunches, spaced all around the 27-kilometer circumference of the collider. Each bunch of up to 100 billion protons will be the size of a needle, just a few centimeters long and squeezed down to 16 microns in diameter (about the same as the thinnest of human hairs) at the collision points. At four locations around the ring, these needles will pass through one another, producing more than 600 million particle collisions every second. The collisions, or events, as physicists call them, actually will occur between particles that make up the protons—quarks and gluons. The most cataclysmic of the smashups will release about a seventh of the energy available in the parent protons, or about 2 TeV. (For the same reason, the Tevatron falls short of exploring terascale physics by about a factor of five, despite the 1-TeV energy of its protons and antiprotons.)

Four giant detectors—the largest would roughly half-fill the Notre Dame cathedral in Paris, and the heaviest contains more iron than the Eiffel Tower—will track and measure the thousands of particles spewed out by each collision occurring at their centers. Despite the detectors' vast size, some elements of them must be positioned with a precision of 50 microns.

The nearly 100 million channels of data streaming from each of the two largest detectors would fill 100,000 CDs every second, enough to produce a stack to the moon in six months. So instead of attempting to record it all, the experiments will have what are called trigger and data-acquisition systems, which act like vast spam filters, immediately discarding almost all the information and sending the data from only the most promising-looking 100 events each second to the LHC's central computing system at CERN, the European laboratory for particle physics and the collider's home, for archiving and later analysis.

A "farm" of a few thousand computers at CERN will turn the filtered raw data into more compact data sets organized for physicists to comb through. Their analyses will take place on a so-called grid network comprising tens of thousands of PCs at institutes around the world, all connected to a hub of a dozen major centers on three continents that are in turn linked to CERN by dedicated optical cables.

JOURNEY OF A THOUSAND STEPS

In the coming months, all eyes will be on the accelerator. The final connections between adjacent magnets in the ring were made in early November, and as we go to press in mid-December one of the eight sectors has been cooled almost to the cryogenic temperature required for operation, and the cooling of a second has begun. One sector was cooled, powered up and then returned to room temperature earlier in 2007. After the operation of the sectors has been tested, first individually and then together as an integrated system, a beam of protons will be injected into one of the two beam pipes that carry them around the machine's 27 kilometers.

The series of smaller accelerators that supply the beam to the main LHC ring has already been checked out, bringing protons with an energy of 0.45 TeV "to the doorstep" of where they will be injected into the LHC. The first injection of the beam will be a critical step, and the LHC scientists will start with a low-intensity beam to reduce the risk of damaging LHC hardware. Only when they have carefully assessed how that "pilot" beam responds inside the LHC and have made fine corrections to the steering magnetic fields will they proceed to higher intensities. For the first running at the design energy of 7 TeV, only a single bunch of protons will circulate in each direction instead of the nearly 3,000 that constitute the ultimate goal.

As the full commissioning of the accelerator proceeds in this measured step-by-step fashion, problems are sure to arise. The big unknown is how long the engineers and scientists will take to overcome each challenge. If a sector has to be brought back to room temperature for repairs, it will add months.

The four experiments—ATLAS, ALICE, CMS and LHCb—also have a lengthy process of completion ahead of them, and they must be closed up before the beam commissioning begins. Some extremely fragile units are still being installed, such as the so-called vertex locator detector that was positioned in LHCb in mid-November. During my visit, as one who specialized in theoretical rather than experimental physics many years ago in graduate school, I was struck by the thick rivers of thousands of cables required to carry all the channels of data from the detectors—every cable individually labeled and needing to be painstakingly matched up to the correct socket and tested by present-day students.

Although colliding beams are still months in the future, some of the students and postdocs already have their hands on real data, courtesy of cosmic rays sleeting down through the Franco-Swiss rock and passing through their detectors sporadically. Seeing how the detectors respond to these interlopers provides an important reality check that everything is working together correctly—from the voltage supplies to the detector elements themselves to the electronics of the readouts to the data-acquisition software that integrates the millions of individual signals into a coherent description of an "event."

ALL TOGETHER NOW

When everything is working together, including the beams colliding at the center of each detector, the task faced by the detectors and the data-processing systems will be Herculean. At the design luminosity, as many as 20 events will occur with each crossing of the needlelike bunches of protons. A mere 25 nanoseconds pass between one crossing and the next (some have larger gaps). Product particles sprayed out from the collisions of one crossing will still be moving through the outer layers of a detector when the next crossing is already taking place. Individual elements in each of the detector layers respond as a particle of the right kind passes through it. The millions of channels of data streaming away from the detector produce about a megabyte of data from each event: a petabyte, or a billion megabytes, of it every two seconds.

The trigger system that will reduce this flood of data to manageable proportions has multiple levels. The first level will receive and analyze data from only a subset of all the detector's components, from which it can pick out promising events based on isolated factors such as whether an energetic muon was spotted flying out at a large angle from the beam axis. This so-called level-one triggering will be conducted by hundreds of dedicated computer boards—the logic embodied in the hardware. They will select 100,000 bunches of data per second for more careful analysis by the next stage, the higher-level trigger.

The higher-level trigger, in contrast, will receive data from all of the detector's millions of channels. Its software will run on a farm of computers, and with an average of 10 microseconds elapsing between each bunch approved by the level-one trigger, it will have enough time to "reconstruct" each event. In other words, it will project tracks back to common points of origin and thereby form a coherent set of data—energies, momenta, trajectories, and so on—for the particles produced by each event.

The higher-level trigger passes about 100 events per second to the hub of the LHC's global network of computing resources—the LHC Computing Grid. A grid system combines the processing power of a network of computing centers and makes it available to users who may log in to the grid from their home institutes.

The LHC's grid is organized into tiers. Tier 0 is at CERN itself and consists in large part of thousands of commercially bought computer processors, both PC-style boxes and, more recently, "blade" systems similar in dimensions to a pizza box but in stylish black, stacked in row after row of shelves. Computers are still being purchased and added to the system. Much like a home user, the people in charge look for the ever moving sweet spot of most bang for the buck, avoiding the newest and most powerful models in favor of more economical options.

The data passed to Tier 0 by the four LHC experiments' data-acquisition systems will be archived on magnetic tape. That may sound old-fashioned and low-tech in this age of DVD-RAM disks and flash drives, but François Grey of the CERN Computing Center says it turns out to be the most cost-effective and secure approach.

Tier 0 will distribute the data to the 12 Tier 1 centers, which are located at CERN itself and at 11 other major institutes around the world, including Fermilab and Brookhaven National Laboratory in the U.S., as well as centers in Europe, Asia and Canada. Thus, the unprocessed data will exist in two copies, one at CERN and one divided up around the world. Each of the Tier 1 centers will also host a complete set of the data in a compact form structured for physicists to carry out many of their analyses.

The full LHC Computing Grid also has Tier 2 centers, which are smaller computing centers at universities and research institutes. Computers at these centers will supply distributed processing power to the entire grid for the data analyses.

ROCKY ROAD

With all the novel technologies being prepared to come online, it is not surprising that the LHC has experienced some hiccups—and some more serious setbacks—along the way. Last March a magnet of the kind used to focus the proton beams just ahead of a collision point (called a quadrupole magnet) suffered a "serious failure" during a test of its ability to stand up against the kind of significant forces that could occur if, for instance, the magnet's coils lost their superconductivity during operation of the beam (a mishap called quenching). Part of the supports of the magnet had collapsed under the pressure of the test, producing a loud bang like an explosion and releasing helium gas. (Incidentally, when workers or visiting journalists go into the tunnel, they carry small emergency breathing apparatuses as a safety precaution.)

These magnets come in groups of three, to squeeze the beam first from side to side, then in the vertical direction, and finally again side to side, a sequence that brings the beam to a sharp focus. The LHC uses 24 of them, one triplet on each side of the four interaction points. At first the LHC scientists did not know if all 24 would need to be removed from the machine and brought aboveground for modification, a time-consuming procedure that could have added weeks to the schedule. The problem was a design flaw: the magnet designers (researchers at Fermilab) had failed to take account of all the kinds of forces the magnets had to withstand. CERN and Fermilab researchers worked feverishly, identifying the problem and coming up with a strategy to fix the undamaged magnets in the accelerator tunnel. (The triplet damaged in the test was moved aboveground for its repairs.)

In June, CERN director general Robert Aymar announced that because of the magnet failure, along with an accumulation of minor problems, he had to postpone the scheduled start-up of the accelerator from November 2007 to spring of this year. The beam energy is to be ramped up faster to try to stay on schedule for "doing physics" by July.

Although some workers on the detectors hinted to me that they were happy to have more time, the seemingly ever receding start-up date is a concern because the longer the LHC takes to begin producing sizable quantities of data, the more opportunity the Tevatron has—it is still running—to scoop it. The Tevatron could find evidence of the Higgs boson or something equally exciting if nature has played a cruel trick and given it just enough mass for it to show up only now in Fermilab's growing mountain of data.

Holdups also can cause personal woes through the price individual students and scientists pay as they delay stages of their careers waiting for data.

Another potentially serious problem came to light in September, when engineers discovered that sliding copper fingers inside the beam pipes known as plug-in modules had crumpled after a sector of the accelerator had been cooled to the cryogenic temperatures required for operation and then warmed back to room temperature.

At first the extent of the problem was unknown. The full sector where the cooling test had been conducted has 366 plug-in modules, and opening up every one for inspection and possibly repair would have been terrible. Instead the team addressing the issue devised a scheme to insert a ball slightly smaller than a Ping-Pong ball into the beam pipe—just small enough to fit and be blown along the pipe with compressed air and large enough to be stopped at a deformed module. The sphere contained a radio transmitting at 40 megahertz—the same frequency at which bunches of protons will travel along the pipe when the accelerator is running at full capacity—enabling the tracking of its progress by beam sensors that are installed every 50 meters. To everyone's relief, this procedure revealed that only six of the sector's modules had malfunctioned, a manageable number to open up and repair.

When the last of the connections between accelerating magnets was made in November, completing the circle and clearing the way to start cooling down all the sectors, project leader Lyn Evans commented, "For a machine of this complexity, things are going remarkably smoothly, and we're all looking forward to doing physics with the LHC next summer."

CHAPTER 14
THE STARS

The Sun and other stars use nuclear fusion reactions to convert mass into energy. Eventually, when a star's nuclear fuel is depleted, the star must burn out.

INTRODUCTION

THE MOST ABUNDANT objects in space visible from Earth are stars—hundreds of billions of stars in our Milky Way galaxy alone. The following two excerpts relate key discoveries regarding the distance to stars and their nuclear source of energy.

Friedrich Wilhelm Bessel (1784–1846) was born in Germany and began life as an accountant in a shipping firm. This led him to an interest in navigation and, from there, to astronomy. His refined calculations of the orbit of Halley's Comet brought him to the attention of the astronomical community, and he rose from an assistant to the director of the Königsberg Observatory in a few years. This paper reports on the first measurement of the distance to a star, and hence on humanity's first realization of the true size of the universe. The method, which astronomers call "parallax" and most people call "triangulation," involves measuring the angles of the lines of sight to the star from different places on the Earth's orbit, and then using geometry to deduce the distance to the star. Until the early eighteenth century, telescopes simply weren't good enough to make these sorts of measurements, and Bessel won a race with several other observatories to be the first to do so.

The Parallax of 61 Cygni
by Friedrich Wilhelm Bessel (1838)

ESTEEMED SIR,—HAVING SUCCEEDED in obtaining a long-looked-for result, and presuming that it will interest so great and zealous an explorer of the heavens as yourself, I take the liberty of making a communication to you thereupon. Should you consider this communication of sufficient importance to lay before other friends of Astronomy, I not only have no objection, but request you to do so. With this view, I might have sent it to you through Mr. Baily; and I should have preferred this course, as it would have interfered less with the important affairs claiming your immediate attention on your return to England. But, to you, I can write in my own language, and thus secure my meaning from indistinctness.

After so many unsuccessful attempts to determine the parallax of a fixed star, I thought it worth while to try what might be accomplished by means of the accuracy which my great Fraunhofer Heliometer [*a new telescope*] gives to the observations. I undertook to make this investigation upon the star 61 *Cygni*, which, by reason of its great proper motion, is perhaps the best of all; which affords the advantage of being a double star, and on that account may be observed with, greater accuracy; and which is so near the pole that, with the exception of a small part of the year, it can always be observed at night at a sufficient distance from the horizon. I began the comparisons of this star in September, 1834, by measuring its distance from two small stars of the 11th magnitude, of which one precedes, and the other is to the northward. But I soon perceived that the atmosphere was seldom sufficiently favourable to allow of the observation of stars so small; and, therefore, I resolved to select brighter ones, although somewhat more distant. In the year 1835, researches on the length of the pendulum at Berlin took me away for three months from the observatory; and when I returned, Halley's Comet had made its appearance, and claimed all the clear nights. But, in 1837 these obstacles were removed.

I selected among the small stars which surround that double star, two between the 9th and 10th magnitudes; of which one, a, is nearly perpendicular to the line of direction of the double star; the other, .6, nearly in this direction, I have measured with the heliometer the distances of these stars from the point which bisects the distance between the two stars of 61 *Cygni*; as I considered this kind of observation the most correct that could be obtained, I have commonly repeated the observations sixteen times every night. When the atmosphere has been unusually unsteady, I have, however, made more numerous repetitions; although, by this, I fear the result has not attained that precision which it would have possessed by fewer observations on more favourable nights. This unsteadiness of the atmosphere is the great obstacle which attaches to all the more delicate astronomical observations. In an unfavourable climate we cannot avoid its prejudicial influence, unless by observing only on the finest nights; by which, however, it would become still more difficult to collect the number of observations necessary for an investigation. …

The tables contain all my measures of distance, freed from the effects of refraction and aberration, and reduced to the beginning of 1838. I have employed the preceding list of the observations of the distances of the star 61 *Cygni* from *a* and *b*, in two different ways, in order to deduce from it results for the annual parallax of *a Cygni*. I have first assumed *a"* and *b"* as independent of each other; or, in other words, considered it as not improbable that *a* and *b* themselves may possess sensible parallax. In this way I have found,

For the star *a*   Mean error
Mean distance for the beginning of 1838…   461"-6094
Annual variation = +4"-3915 - 0".0543…   +4 -3372   ± 0".0398   '
Difference of annual parallax of 61 and a… *a*" — +0 .3690   ±0 .0283

For the star *b*
Mean distance for the beginning of 1838…   706-2909
Annual variation = -2".825 + 0".2426…   -2-5824   ±0 -0434
Difference of annual parallax of 61 and b… β" = +0.2605   ±0 .0278

As the mean error of the annual parallax of 61 Cygni (= 0".3136) is only ±0".02020, and consequently not 1/15 of its value computed; and these comparisons show that the progress of the influence of the parallax, which the observations indicate, follows the theory as nearly as can be expected considering its smallness, we can no longer doubt that this parallax is sensible. Assuming it 0".3136, we find the distance of the star 61 Cygni from the sun 657,700 mean distances of the earth from the sun: light employs 10.3 years to traverse this distance.

Stellar Energy
by Henry Norris Russell (1939)

HANS BETHE (1906–2005) was born in Strasburg and educated at Frankfort and Munich, where he received his doctorate in theoretical physics. He left his native Germany is 1933 after the Nazis came to power and eventually joined the faculty at Cornell University, where he remained for the rest of his career. During World War II he was prominently involved in the Manhattan Project, and after the war was an advisor to the American government on many matters, playing a key role in the development of the Nuclear Test Ban Treaty. He received the Nobel Prize for the work described here in 1967.

As is explained in the reading, his revolutionary work on the energy sources of stars, eventually published in 1939, was so unexpected that the organizers of a major conference couldn't put him on the program. Instead, they asked Princeton astronomer Henry Norris Russell to summarize the work, which he did.

One word of caution: at the time, astronomers overestimated the temperature in the interior of stars. Consequently, the precise reactions outlined by Bethe are now known to occur only in stars larger than the sun. Nevertheless, the idea that stars derive their energy from nuclear fusion remains a central premise of modern astronomy.

* * *

This lecture should begin by an apology to the audience. The speaker tonight should be the man whose recent and brilliant work has inaugurated a new and very promising stage of astrophysical study—Professor H. A. Bethe of Cornell. The planners of the program can plead but one excuse—his results had not been announced when it was prepared. They are not yet published in detail; but Professor Bethe has most generously given me a full copy of the proof of his forthcoming paper, and amplified this in personal conversation. What there is of novelty in the present discussion should be credited entirely to him, and not to myself. Indeed my report, apart from this, is mainly upon the work of others in a field in which I did a little pioneering a good many years ago.

A century of precise measurement has furnished astronomers with a large accumulated capital of facts about the stars. One of the most remarkable things that is thus revealed is that the stars differ greatly among themselves in some properties, and relatively little in others. For example, the luminosity—the rate of radiation of energy—ranges (roughly) from a million times the Sun's luminosity to a thousandth part of it a ratio of a billion to one. The diameters run from at least 1000 times the Sun's down to 1/30 or less, a range of thirty thousand-fold. For the masses, the range is roughly from 100 times the Sun to 1/10 of the Sun, a thousand-fold. But surface temperatures more than ten times the Sun's or less than 1/4 that of the Sun are practically unheard-of, so that the known range is only forty-fold.

Why should such enormous diversities exist? The limitation of our study is set, on one side, by the obvious fact that we cannot see a star unless it shines—that is, unless it gives out enough light to be perceptible at stellar distances. This obviously sets a limit to our knowledge of *faint* stars, and of *small* stars—which, other things being equal, will shine less brightly because they have fewer square miles of surface, and still more to the study of *cool* stars (of low surface temperature). The heat radiation per square mile varies as the fourth power of the temperature, and a forty-fold increase in the latter changes the former by a factor of about two and a half million. Moreover, at low temperatures, most of the radiant energy is in long infra-red waves, invisible to the eye, so that a star—like a mass of hot iron—ceases to be visible in the dark, long before it stops giving out heat.

But there is no such limitation on the side of great brightness. If we fail to find still brighter stars—though we are fishing for these in very wide waters—it must be because there are no such fish in our sea. Nature herself must in some way set a limit to the brightness attainable by a star—at least as a permanent affair for we know that the short lived outbursts of supernovae are enormously more intense. Such a limitation would explain why stars of very large diameter are never found to be very hot—if they were, they would give out too much light; but it has no obvious relation to the mass of a star.

A notable advance was made more than twenty years ago when it was found observationally (first by Halm) that there is a close relation between the mass of a star and its luminosity.

All later work has confirmed this, and it now appears that, over almost if not quite the whole available range, the total heat-radiation of a star is nearly proportional to the fourth power of its mass. The most remarkable feature is that, for a given mass, the luminosity does not depend much upon the size of the star—the large ones are cool, and give out less heat per square mile, the small ones hot, and give out more; but the net product is nearly the same.

Eddington's explanation of this, completed fifteen years ago, marks the first, and, even now, the greatest success of modern atomic physics in interpreting the stars.

Since the work of Lane in 1870, it has been realized that, if the familiar laws of perfect gases could be applied to the interior of the Sun, the central temperature must be many millions of degrees.

Fortunately for us theorists, the situation is simpler inside the stars; for the outsides of the atoms—the electrons which surround the nucleus—are pretty well knocked off. A hydrogen atom (with one electron) is broken into 2 particles, a helium atom into 3, one of oxygen into 9, and one of iron into 27 (or perhaps 25, if the two tightest bound electrons stay with the nucleus).

The average mass of a free particle (which is what counts) then comes out 1/2 for hydrogen, 4/3 for helium and nearly 2 for all heavier atoms. So we can conclude that the central temperature of the Sun is close to 10 million degrees if it is pure hydrogen, 27 million if it is all helium, and 40 million if it is all composed of heavy atoms—and the result for any assumed mixture can easily be computed.

The temperature of the Sun's surface is but a few thousand degrees. Heat will flow outward from the hot interior "down grade" and the surface will adjust its temperature so that it loses by radiation into space just the amount which it gains by transmission from the deep interior. The rate of this transmission thus determines how bright and hot the Sun (or any other star) will be; it takes place mainly by the transfer of radiation from atom to atom through the gas; and the net opacity, which determines the rate of transfer, may be calculated with considerable accuracy.

When the net flow of heat out to a star's surface is thus calculated, it is found that it depends relatively little on the "model," and not much upon the radius, but changes rapidly, with the mass—at very nearly the rate indicated by observation.

The mass-luminosity relation is thus explained—and with it the fact that stars of less than one-tenth the Sun's have not so far been detected—they are presumably too faint to be seen; and our failure to discover very large masses is connected with the absence of stars of very great luminosity.

When first the Sun's heat was measured, it was realized at once that it could hardly be maintained, even for the duration of history, by ordinary chemical combinations—what we now describe as reactions between whole atoms. A far larger source of energy is found in a slow contraction, compressing the gas and turning the gravitational potential. energy into heat, and this explanation, due to Helmholtz and Kelvin, was still accepted forty years ago.

The energy-supply thus available for the Sun's past history is about thirty million times its present annual expenditure (depending somewhat on the assumed model). But at least half this must still be stored inside the Sun in the form of heat, so that less than fifteen million years of sunshine can be accounted for.

But the Sun has been shining and warming the Earth, very much as now, throughout geological history, which is certainly a hundred times as long as this. There is only one place in the known universe to look for so vast a store of energy—and that is in the minute nuclei of atoms. From these nuclei is liberated the energy of radio-activity—which is of the required order of magnitude, though rather small. But radioactivity itself will not meet our needs, for it takes place at a rate uninfluenced by conditions external to the nucleus—such as the surrounding, temperature. To account for the observed variation of luminosity with mass, we would have to assume that in some inscrutable way the amount of active material had been exactly proportional "in the beginning" to the mass of each star, so as to provide the proper heat supply to maintain its radiation; and also that practically none of it had been allowed to enter the Earth or the other planets—for otherwise their surfaces would be red-hot or even white-hot. It is necessary, therefore, to assume that the great store of energy upon which the stars draw becomes available only under stellar conditions—that is, obviously, at temperatures of millions of degrees. ...

One other thing was clearly understood, twenty years or so ago; the reactions which provide the stars with their energy must be accompanied by a perceptible loss of mass. The theory of special relativity indicated that mass and energy should be interconvertible, at the rate of c^2 units of energy (ergs) for one unit of mass (gram), where c is the velocity of light. The Sun's rate of loss of energy by radiation (3.8×10^{33} ergs per second) is far too great to be comprehensible in ordinary terms; but, if measured in mass-units, it means that the Sun is actually losing 4,200,000 *tons* of energy every second! The Sun's mass (2×10^{33} g) is so great that even at this enormous rate, it would take 15 billion years to reduce it by one part in a thousand.

The nature of the process—or, at least, of one process—by which this transformation of mass into energy can take place, is now well understood, since it can be made to happen in the laboratory, in many different ways. This is the transformation of atoms of one sort into another, with the appearance of an amount of energy corresponding to the change in total mass; and it is to this that we now look as the source of stellar energy.

The masses of the lighter atoms (including their outer electrons) are given in Table 1. They can be very accurately determined—mainly with the mass-spectrograph. The various isotopes—atoms of different weight, but the same chemical properties—are of course listed individually.

From the table it appears that if four hydrogen atoms could in any way be converted into one helium atom, a loss of mass of 0.02866 unit would result. This is 1/141 of the original mass. If, therefore, a thousandth part of the Sun's mass were hydrogen, and could be transmuted into helium, energy enough would be released to keep the Sun shining for 106 million-years.

The excess of the tabulated mass above an even value evidently represents the possible value of the given atom as a source of energy. This excess is fairly considerable for most of the lighter atoms; but oxygen, carbon and helium are relatively very stable. Some atoms heavier than oxygen are still further "down"; but the differences are small and need not concern us here. Per unit of mass, hydrogen is much the best source of energy.

TABLE I: Masses of Light Atoms *

Name	Symbol	Charge	Mass
Electron	E	-1	0.00055
Neutron	N	0	1.00893
Hydrogen	H	1	1.00813
Deuterium	H^2	1	2.01473
Helium	He^3	2	3.01699
	He^4	2	4.00386
Lithium	Li^6	3	6.01686
	Li^7	3	7.01818
Beryllium	Be^9	4	9.01504
Boron	B^{10}	5	10.01631
	B^{11}	5	11.01292
Carbon	C^{12}	6	12.00398
	C^{18}	6	13.00761
Nitrogen	N^{14}	7	14.00750
	N^{15}	7	15.00489
Oxygen	O^{16}	8	16.00000

* These values are from Bethe's computations, with some changes on the basis of his latest work.

Two types of nuclear reaction, in the laboratory, give rise to reactions of the sort considered—the penetration into a heavier nucleus of a neutron, or of a charged particle (proton, deuteron, or alpha-particle). The former are the easiest to produce artificially—since the other nucleus does not repel the neutron. They occur in great variety, and liberate large amounts of energy. For this very reason, it appears that they are not of much astrophysical importance. They happen *too* easily. If there were a lot of neutrons in the interior of a star, they would all collide with atomic nuclei of one sort or another—sometimes building up a heavier atom, sometimes causing the emission of some other particle—and they would be used up in the process, in the twinkling of an eye. There might be an almost explosive outburst of heat (under the altogether unnatural conditions which we have imagined) but anyhow the neutrons would be gone.

We are left, then, as a steady source of energy, with the penetration of one charged nucleus into another. The repulsion between the two is always so great that only an exceptionally violent head-on collision between particles moving much faster than the average would be effective. The probability of such a collision has been calculated. It diminishes very rapidly with the charges of the particles, so that a proton is vastly more likely to succeed than an alpha-particle, while, even so, it stands very little chance of getting into a nucleus with a larger charge than oxygen (provided, at least, that the temperature is not more than twenty million degrees).

What happens after the proton gets in depends on the individual peculiarities of the nucleus which it has hit. It may simply go in and stay—producing a new nucleus greater by one in both charge and mass; or a positive electron may be ejected, leaving a nucleus of the same charge as before but of greater mass; or the old nucleus may break up into two or more pieces—one of which is usually an alpha-particle. None of these changes can happen if the hypothetically resulting nucleus is heavier than the reacting particles, for a quite impossible amount of energy would have to be supplied from no known source. If the mass of the product is less than that of the two particles, the excess energy will appear as kinetic energy when the nucleus breaks up, or an electron is ejected; in the first case, it appears as radiation—a gamma-ray.

All three processes occur in the stars.

The one stable thing in this microcosm of change is the alpha particle (He^4). Apparently nothing can happen to it. The nuclei which might imaginably be formed by collision with a proton (He^5 or Li^5) are unstable, and do not exist at all; a collision of two alpha-particles would give Be^8, which is slightly unstable, and breaks up again. Three fast-moving alpha-particles, colliding simultaneously, might possibly form C^{12} with liberation of energy; but such an event is so very improbable that it can be neglected, except at much higher temperatures.

If, then, hydrogen is "the fuel of the stars" helium is the ashes.

Second. The other light nuclei, up to and including boron, are highly-susceptible to proton collisions. The different ones go through various transformations, but every sequence ends irrevocably in helium—for example, Li^7 4- H = 2 He^4, B^{11} + H = 3 He^4. (In this last case the nucleus breaks into three pieces—all helium.) Hence these elements, if originally present in a star, would be successively exhausted as the reactions went on.

Third. At a temperature somewhat above fifteen million degrees, carbon begins to be attacked, and something quite new happens, which is tabulated by Bethe as follows:

$$C^{12} + H = N^{13} + \gamma$$ 2,500,000 years[2]
$$N^{13} = C^{13} + \epsilon^{+}$$ 9.9 minutes
$$C^{13} + H = N^{14} + \gamma$$ 50,000 years
$$N^{14} + H = O^{15} + \gamma$$ 4,000,000 years
$$O^{15} = N^{15} + \epsilon^{+}$$ 2.1 minutes
$$N^{15} + H = C^{12} + He^{4}$$ 20 years

The proton goes into C^{12} and builds up N^{13} (the spare energy escaping as a γ-ray). This is one of the artificial radio-active elements, which has been produced in the laboratory. It emits a positive electron (ϵ^{+}) and goes over into the carbon isotope C^{13}. The next proton turns this into ordinary nitrogen N^{14}, and another produces a radio-active oxygen O^{15} which goes over into "heavy" nitrogen N^{15}. The average length of time during which an atom of each kind may be expected to last before its next adventure, is given in the table. The $N^{14} + H$ reaction is the slowest; it will then be the "bottle-neck" for the whole process. Bethe's calculations show that it would supply energy enough for the Sun at a temperature of 18,300,000° (assuming 35 percent of hydrogen and 10 percent of nitrogen). As the rate of reaction increases as the 18th power of the temperature, it would do the work, with 1 percent of nitrogen, at a temperature of 20,800,000°. Now these temperatures, which are calculated from pure nuclear theory, agree excellently with the central temperature of the Sun, as calculated from astrophysical data. An almost equally good agreement is found for Sirius (22 million from theory, 26 from observation), and for the very hot and massive star Y Cygni (30 and 32 million). Moreover, the gradual changes in light, diameter, density and surface temperature for stars of different mass along the whole main sequence are in excellent agreement with the results of the theory.

It appears, then, that one great part of the problem of the source of stellar energy has been fully solved. The main sequence is explained in detail. ...

1 Much of the energy in this case is believed to be carried off by that elusive particle, a neutrino.

CHAPTER 15
COSMOLOGY

*The universe began billions of years ago in the big bang,
and it has been expanding ever since.*

INTRODUCTION

T HE ORIGIN AND future of the Cosmos are among the most profound unanswered questions in science. Here we reproduce two key contributions that point to a universe with a specific date of birth almost 14 billion years ago.

Henrietta Leavitt (1868–1921) was educated at what is now Radcliffe College, where she studied astronomy. She started as an unpaid volunteer at Harvard Observatory, analyzing photographic plates, but quickly rose to the position of chief of the photographic photometry department. In this paper, appended to a publication of Harvard College Observatory Director Edward Charles Pickering (1846–1919), she established the fact that period of Cepheid variable stars is related to their brightness, a crucial fact needed to establish astronomical distances. Her work was the foundation used later by Edwin Hubble to establish the expansion of the universe.

Periods of 25 Variable Stars in the Small Magellanic Cloud

by Henrietta Swan Leavitt (1912)

THE FOLLOWING STATEMENT regarding the periods of 25 variable stars in the Small Magellanic Cloud has been prepared by Miss Leavitt.

A Catalogue of 1777 variable stars in the two Magellanic Clouds is given in H.A. 60, No. 4. The measurement and discussion of these objects present problems of unusual difficulty, on account of the large area covered by the two regions, the extremely crowded distribution of the stars contained in them, the faintness of the variables, and the shortness of their periods. As many of them never become brighter than the fifteenth magnitude, while very few exceed the thirteenth magnitude at maximum, long exposures are necessary, and the number of available photographs is small. The determination of absolute magnitudes for widely separated sequences of comparison stars of this degree of faintness may not be satisfactorily completed for some time to come. With the adoption of an absolute scale of magnitudes for stars in the North Polar Sequence, however, the way is open for such a determination.

Fifty-nine of the variables in the Small Magellanic Cloud were measured in 1904, using a provisional scale of magnitudes, and the periods of seventeen of them were published in H.A. 60, No. 4, Table VI. They resemble the variables found in globular clusters, diminishing slowly in brightness, remaining near minimum for the greater part of the time, and increasing very rapidly to a brief maximum. Table I gives all the periods which have been determined thus far, 25 in number, arranged in the order of their length. The first five columns contain the Harvard Number, the brightness at maximum and at minimum as read from the light curve, the epoch expressed in days following J.D. 2,410,000, and the length of the period expressed in days. The Harvard Numbers in the first column are placed in Italics, when the period has not been published hitherto. A remarkable relation between the brightness of these variables and the length of their periods will be noticed. In H.A. 60, No. 4, attention was called to the fact that the brighter variables have the longer periods, but at that time it was felt that the number was too small to warrant the drawing of general conclusions. The periods of 8 additional variables which have been determined since that time, however, conform to the same law.

TABLE 1.

PERIODS OF VARIABLE STARS IN THE SMALL MAGELLANIC CLOUD.

H.	Max.	Min.	Epoch.	Period.	Res. M.	Res. m.	H.	Max.	Min.	Epoch.	Period.	Res. M.	Res. m.
1505	14.8	16.1	0.02	1.25336	−0.6	−0.5	1400	14.1	14.8	4.0	6.650	+0.2	−0.3
1436	14.8	16.4	0.02	1.6637	−0.3	+0.1	1355	14.0	14.8	4.8	7.483	+0.2	−0.2
1446	14.8	16.4	1.38	1.7620	−0.3	+0.1	1374	13.9	15.2	6.0	8.397	+0.2	−0.3
1506	15.1	16.3	1.08	1.87502	+0.1	+0.1	818	13.6	14.7	4.0	10.336	0.0	0.0
1413	14.7	15.6	0.35	2.17352	−0.2	−0.5	1610	13.4	14.6	11.0	11.645	0.0	0.0
1460	14.4	15.7	0.00	2.913	−0.3	−0.1	1365	13.8	14.8	9.6	12.417	+0.4	+0.2
1422	14.7	15.9	0.6	3.501	+0.2	+0.2	1351	13.4	14.4	4.0	13.08	+0.1	−0.1
842	14.6	16.1	2.61	4.2897	+0.3	+0.6	827	13.4	14.3	11.6	13.47	+0.1	−0.2
1425	14.3	15.3	2.8	4.547	0.0	−0.1	822	13.0	14.6	13.0	16.75	−0.1	+0.3
1742	14.3	15.5	0.95	4.9866	+0.1	+0.2	823	12.2	14.1	2.9	31.94	−0.3	+0.4
1646	14.4	15.4	4.30	5.311	+0.3	+0.1	824	11.4	12.8	4.	65.8	−0.4	−0.2
1649	14.3	15.2	5.05	5.323	+0.2	−0.1	821	11.2	12.1	97.	127.0	−0.1	−0.4
1492	13.8	14.8	0.6	6.2926	−0.2	−0.4							

The relation is shown graphically in Figure 1, in which the abscissas are equal to the periods, expressed in days, and the ordinates are equal to the corresponding magnitudes at maxima and at minima. The two resulting curves, one for maxima and one for minima, are surprisingly smooth, and of remarkable form. In Figure 2, the abscissas are equal to the logarithms of the periods, and the ordinates to the corresponding magnitudes, as in Figure 1.

A straight line can readily be drawn among each of the two series of points corresponding to maxima and minima, thus showing that there is a simple relation between the brightness of the variables and their periods. The logarithm of the period increases by about 0.48 for each increase of one magnitude in brightness. The residuals of the maximum and minimum of each star from the lines in Figure 2 are given in the sixth and seventh columns of Table I. It is possible that the deviations from a straight line may become smaller when an absolute scale of magnitudes is used, and they may even indicate the corrections that need to be applied to the provisional scale. It should be noticed that the average range, for bright and faint variables alike, is about 1.2 magnitudes. Since the variables are probably at nearly the same distance from the Earth, their periods are apparently associated with their actual emission of light, as determined by their mass, density, and surface brightness.

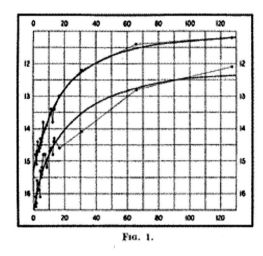

FIG. 1.

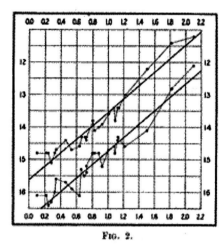

FIG. 2.

The faintness of the variables in the Magellanic Clouds seems to preclude the study of their spectra, with our present facilities. A number of brighter variables have similar light curves, as UY Cygni, and should repay careful study. The class of spectrum ought to be determined for as many such objects as possible. It is to be hoped, also, that the parallaxes of some variables of this type may be measured. Two fundamental questions upon which light may be thrown by such inquiries are whether there are definite limits to the mass of variable stars of the cluster type, and if the spectra of such variables having long periods differ from those of variables whose periods are short.

The facts known with regard to these 25 variables suggest many other questions with regard to distribution, relations to star clusters and nebulae, differences in the forms of the light curves, and the extreme range of the length of the periods. It is hoped that a systematic study of the light changes of all the variables, nearly two thousand in number, in the two Magellanic Clouds may soon be undertaken at this Observatory.

A Relation Between Distance and Radial Velocity Among Extra-Galactic Nebulae

by Edwin Hubble (1929)

EDWIN HUBBLE (1889–1953) was educated at the University of Chicago and Oxford (as a Rhodes Scholar). He received his PhD at Chicago and, after service in World War I, joined the Carnegie Institution of Washington and the staff at the (then) new observatory on Mount Wilson, near Los Angeles. There he discovered the most significant fact in cosmology—the expansion of the universe. In this paper, he announces that discovery, based on his ability to use the Mount Wilson telescope to see individual stars in nearby galaxies. He measured distances by monitoring Cepheid variable stars, whose energy output can be determined from their period of variation, and he measured the speed at which they are receding from us from the red shift of the light emitted by the stars in the galaxy. By demonstrating that the farther away a galaxy is the faster it is moving, Hubble laid the groundwork for the Big Bang.

* * *

Distances of extra-galactic nebulae depend ultimately upon the application of absolute-luminosity criteria to involved stars whose types can be recognized. These include, among others, Cepheid variables, novae, and blue stars involved in emission nebulosity. Numerical values depend upon the zero point of the period-luminosity relation among Cepheids, the other criteria merely check the order of the distances. This method is restricted to the few nebulae which are well resolved by existing instruments. A study of these nebulae, together with those in which any stars at all can be recognized, indicates the probability of an approximately uniform upper limit to the absolute luminosity of stars, in the late-type spirals and irregular nebulae at least, of the order of M (photographic) = -6.3.[1] The apparent luminosities of the brightest stars in such nebulae are thus criteria which, although rough and to be applied with caution, furnish reasonable estimates of the distances of all extra-galactic systems in which even a few stars can be detected.

Finally, the nebulae themselves appear to be of a definite order of absolute luminosity, exhibiting a range of four or five magnitudes about an average value M (visual) = -15.2. The application of this statistical average to individual cases can rarely be used to advantage, but where considerable numbers are involved, and especially in the various clusters of nebulae, mean apparent luminosities of the nebulae themselves offer reliable estimates of the mean distances.

OBJECT	m_3	r^8	V	m_t	M_t
S. Mag.		0.032	+ 170	1.5	-16.0
L. Mag.		0.034	+ 290	0.5	17.2
N. G. C. 6822		0.214	- 130	9.0	12.7
598		0.263	- 70	7.0	15.1
221		0.275	- 185	8.8	13.4
224		0.275	- 220	5.0	17.2
5457	17.0	0.45	+ 200	9.9	13.3
4736	17.3	0.5	+ 290	8.4	15.1
5194	17.3	0.5	+ 270	7.4	16.1
4449	17.8	0.63	+ 200	9.5	14.5
4214	18.3	0.8	+ 300	11.3	13.2
3031	18.5	0.9	- 30	8.3	16.4
3627	18.5	0.9	+ 650	9.1	15.7
4826	18.5	0.9	+ 150	9.0	15.7
5236	18.5	0.9	+ 500	10.4	14.4
1068	18.7	1.0	+ 920	9.1	15.9
5055	19.0	1.1	+ 450	9.6	15.6
7331	19.0	1.1	+ 500	10.4	14.8
4258	19.5	1.4	+ 500	8.7	17.0
4151	20.0	1.7	+ 960	12.0	14.2
4382		2.0	+ 500	10.0	16.5
4472		2.0	+ 850	8.8	17.7
4486		2.0	+ 800	9.7	16.8
4649		2.0	+ 1090	9.5	17.0
Mean					-15.5

m_s = photographic magnitude of brightest stars involved.

r = distance in units of 10^6 parsecs. The first two are Shapley's values.

v = measured velocities in km./sec. N. G. C. 6822, 221, 224 and 5457 are recent determinations by Humason.

m_t = Holetschek's visual magnitude as corrected by Hopmann. The first three objects were not measured by Holetschek, and the values of m_t represent estimates by the author based upon such data as are available.

M_t = total visual absolute magnitude computed from m_t and r.

Radial velocities of 46 extra-galactic nebulae are now available, but individual distances are estimated for only 24. For one other, N. G. C. 3521, an estimate could probably be made, but no photographs are available at Mount Wilson. The data are given in table 1. The first seven distances are the most reliable, depending, except for M 32 the companion of M 31, upon extensive investigations of many stars involved. The next thirteen distances, depending upon the criterion of a uniform upper limit of stellar luminosity, are subject to considerable probable errors but are believed to be the most reasonable values at present available. The last four objects appear to be in the Virgo Cluster. The distance assigned to the cluster, 2×10^6 parsecs, is derived from the distribution of nebular luminosities, together with luminosities of stars in some of the later-type spirals, and differs somewhat from the Harvard estimate of ten million light years.

The data in the table indicate a linear correlation between distances and velocities, whether the latter are used directly or corrected for solar motion, according to the older solutions. …

The 22 nebulae for which distances are not available can be treated in two ways. First, the mean distance of the group derived from the mean apparent magnitudes can be compared with the mean of the velocities corrected for solar motion. The result, 745 km./sec. for a distance of 1.4×10^6 parsecs, falls between the two previous solutions and indicates a value for K of 530 as against the proposed value, 500 km./sec.

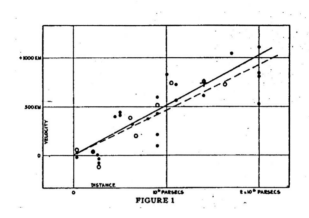

FIGURE 1

FIGURE 1
VELOCITY-DISTANCE RELATION AMONG EXTRA-GALACTIC NEBULAE.

Radial velocities, corrected for solar motion, are plotted against distances estimated from involved stars and mean luminosities of nebulae in a cluster. The black discs and full line represent the solution for solar motion using the nebulae individually; the circles and broken line represent the solution combining the nebulae into groups; the cross represents the mean velocity corresponding to the mean distance of 22 nebulae whose distances could not be estimated individually.

Secondly, the scatter of the individual nebulae can be examined by assuming the relation between distances and velocities as previously determined. Distances can then be calculated from the velocities corrected for solar motion, and absolute magnitudes can be derived from the apparent magnitudes. The results are given in table 2 and may be compared with the distribution of absolute magnitudes among the nebulae in table 1, whose distances are derived from other criteria. N. G. C. 404 can be excluded, since the observed velocity is so small that the peculiar motion must be large in comparison with the distance effect. The object is not necessarily an exception, however, since a distance can be assigned for which the peculiar motion and the absolute magnitude are both within the range previously determined. The two mean magnitudes, −15.3 and −15.5, the ranges, 4.9 and 5.0 mag., and the frequency distributions are closely similar for these two entirely independent sets of data; and even the slight difference in mean magnitudes can be attributed to the selected, very bright, nebulae in the Virgo Cluster. This entirely unforced agreement supports the validity of the velocity-distance relation in a very evident matter. Finally, it is worth recording that the frequency distribution of absolute magnitudes in the two tables combined is comparable with those found in the various clusters of nebulae.

The results establish a roughly linear relation between velocities and distances among nebulae for which velocities have been previously published, and the relation appears to dominate the distribution of velocities. In order to investigate the matter on a much larger scale, Mr. Humason at Mount Wilson has initiated a program of determining velocities of the most distant nebulae that can be observed with confidence. These, naturally, are the brightest nebulae in clusters of nebulae. The first definite result, $v = +3779$ km./sec. for N. G. C. 7619, is thoroughly consistent with the present conclusions. Corrected for the solar motion, this velocity is +3910, which, with $K = 500$, corresponds to a distance of 7.8×10^6 parsecs. Since the apparent magnitude is 11.8, the absolute magnitude at such a distance is −17.65, which is of the right order for the brightest nebulae in a cluster. A preliminary distance, derived independently from the cluster of which this nebula appears to be a member, is of the order of 7×10^6 parsecs.

New data to be expected in the near future may modify the significance of the present investigation or, if confirmatory, will lead to a solution having many times the weight. For this reason it is thought premature to discuss in detail the obvious consequences of the present results. For example, if the solar motion with respect to the clusters represents the rotation of the galactic system, this motion could be subtracted from the results for the nebulae and the remainder would represent the motion of the galactic system with respect to the extra-galactic nebulae.

The outstanding feature, however, is the possibility that the velocity-distance relation may represent the de Sitter effect, and hence that numerical data may be introduced into discussions of the general curvature of space. In the de Sitter cosmology, displacements of the spectra arise from two sources, an apparent slowing down of atomic vibrations and a general tendency of material particles to scatter. The latter involves an acceleration and hence introduces the element of time. The relative importance of these two effects should determine the form of the relation between distances and observed velocities; and in this connection it may be emphasized that the linear relation found in the present discussion is a first approximation representing a restricted range in distance.

CHAPTER 16
THE EARTH AND OTHER PLANETS

The Earth, one of eight planets that orbit the Sun, formed
4.5 billion years ago from a great cloud of dust.

INTRODUCTION

T HE ORIGINS AND nature of the planets was a central focus of astronomical research in the eighteenth century. The following two excerpts, both building on the work of Isaac Newton, present theoretical and experimental approaches to planetary sciences from the 1790s.

Pierre Simon, Marquis de Laplace (1749–1827), French physicist and mathematician, was celebrated as the "French Newton" for his contributions to celestial mechanics. At a time when other scientists thought the solar system would collapse without divine intervention, he showed that Newton's laws were sufficient to yield a stable, albeit dynamic, solar system. In this excerpt from Book V of his epic *The System of the World*, Laplace outlines his nebular hypothesis, which has become the accepted theory of solar system formation from a rotating cloud of dust and gas.

The System of the World[1]
by Pierre Simon Laplace (1796)

BOOK V. SUMMARY OF THE HISTORY OF ASTRONOMY
Chapter VI. Considerations on the System of the World, and on the Future
Progress of Astronomy

SUCH IS UNQUESTIONABLY the constitution of the solar system. The immense globe of the Sun, the focus of these motions, revolves upon its axis in twenty-five days and

---

1 French translation by Henry H. Harte, Dublin, 1830.

a half. Its surface is covered with an ocean of luminous matter. Beyond it the planets, with their satellites, move, in orbits nearly circular, and in planes little inclined to the ecliptic. Innumerable comets, after having approached the Sun, recede to distances, which evince that his empire extends beyond the known limits of the planetary system. This luminary not only acts by its attraction upon all these globes, and compels them to move around him, but imparts to them both light and heat, his benign influence gives birth to the animals and plants which cover the surface of the Earth, and analogy induces us to believe, that he produces similar effects on the planets; for, it is not natural to supposed that matter, of which we see the fecundity develop itself in such various ways, should be sterile upon such a planet as Jupiter, which, like the Earth, has its days, its nights, and its years, and on which observation discovers changes that indicate very active forces. Man, formed for the temperature which he enjoys upon the Earth, could not, according to all appearance, live upon the other planets; but ought there not to be a diversity of organization suited to the various temperatures of the globes of this universe? If the difference of elements and climates alone causes such variety in the productions of the Earth, how infinitely diversified must be the productions of the planets and their satellites? The most active imagination cannot form any just idea of them, but still their existence is, at least, extremely probable.

However arbitrary the elements of the system of the planets may be, there exists between them some very remarkable relations, which may throw light on their origin. Considering it with attention, we are astonished to see all the planets move round the Sun from west to east, and nearly in the same plane, all the satellites moving round their respective planets in the same direction, and nearly in the same plane with the planets. Lastly, the Sun, the planets, and those satellites in which a motion of rotation have been observed, turn on their own axes, in the same direction, and nearly in the same plane as their motion of projection.

The satellites exhibit in this respect a remarkable peculiarity. Their motion of rotation is exactly equal to their motion of revolution; so that they always present the same hemisphere to their primary. At least, this has been observed for the Moon, for the four satellites of Jupiter, and for the last satellite of Saturn, the only satellites whose rotation has hitherto been recognized.

Phenomena so extraordinary, are not the effect of irregular causes. By subjecting their probability to computation, it is found that there is more than two thousand to one against the hypothesis that they are the effect of chance, which is a probability much greater than that on which most of the events of history, respecting which there does not exist a doubt, depends. We ought therefore to be assured with the same confidence, that a primitive cause has directed the planetary motions.

Another phenomenon of the solar system equally remarkable, is the small excentricity of the orbits of the planets and their satellites, while those of the comets are very much extended. The orbits of this system present no intermediate shades between a great and small excentricity. We are here again compelled to acknowledge the effect of a regular cause; chance alone could not have given a form nearly circular to the orbits of all the planets. It is necessary that the cause which determined the motions of these bodies, rendered them also nearly circular. This cause then must also have influenced the great excentricity of the orbits of comets, and their motion in every direction; for, considering the orbits of retrograde comets, as being inclined more than one hundred degrees, which would be the case if the bodies had been projected at random.

What is this primitive cause? In the concluding note of this work I will suggest an hypothesis which appears to me to result with a great degree of probability, from the preceding phenomena, which however I present with that diffidence, which ought always to attach to whatever is not the result of observation and computation.

Whatever be the true cause, it is certain that the elements of the planetary system are so arranged as to enjoy the greatest possible stability, unless it is deranged by the intervention of foreign causes. From the sole circumstance that the motions of the planets and satellites are performed in orbits nearly circular, in the same direction, and in planes which are inconsiderably inclined to each other, the system will always oscillate about a mean state, from which it will deviate but by very small quantities. The mean motions of rotation and of revolution of these different bodies are uniform, and their mean distances from the foci of the principal forces which actuate them are constant; all the secular inequalities are periodic. …

Note VII and Last

Let us consider whether we can assign the true cause. Whatever may be its nature, since it has produced or influenced the direction of the planetary motions, it must have embraced them all within the sphere of its action and, considering the immense distance which intervenes between them, nothing could have effected this but a fluid of almost indefinite extent. In order to have impressed on them all a motion, circular and in the same direction about the Sun, this fluid must environ this star, like an atmosphere. From a consideration of the planetary motions, we are therefore brought to the conclusion, that in consequence of an excessive heat, the solar atmosphere originally extended beyond the orbits of all the planets, and that it has successively contracted itself within its present limits.

In the primitive state in which we have supposed the Sun to be, it resembles those substances which are termed nebulae, which, when seen through telescopes, appear to be composed of a nucleus, more or less brilliant, surrounded by a nebulosity, which, by condensing on its surface, transforms it into a star. If all the stars are conceived to be similarly formed, we can suppose their anterior state of nebulosity to be preceded by other states, in which the nebulous matter was more or less diffuse, the nucleus being at the same time more or less brilliant. By going back in this manner, we shall arrive at a state of nebulosity so diffuse, that its existence can with difficulty be conceived.

For a considerable time back, the particular arrangement of some stars visible to the naked eye, has engaged the attention of philosophers. Mitchel remarked long since how extremely improbable it was that the stars composing the constellation called the Pleiades, for example, should be confined within the narrow space which contains them, by the sole chance of hazard; from which he inferred that this group of stars, and the similar groups which the heavens present to us, are the effects of a primitive cause, or of a primitive law of nature. These groups are a general result of the condensation of nebulae of several nuclei; for it is evident that the nebulous matter being perpetually attracted by these different nuclei, will in like manner form stars very near to each other, revolving the one about the other like to the double stars, whose respective motions have been already recognized.

But in what manner has the solar atmosphere determined the motions of rotation and revolution of the planets and satellites? If these bodies had penetrated deeply into this atmosphere, its resistance would cause them to fall on the Sun. We may therefore suppose that the planets were formed at its successive limits, by the condensation of zones of vapours, which it must, while it was cooling, have abandoned in the plane of its equator. ...

If all the particles of a ring of vapours continued to condense without separating, they would at length constitute a solid or a liquid ring. But the regularity which this formation requires in all parts of the ring, and in their cooling, ought to make this phenomenon very rare. Thus the solar system presents but one example of it; that of the rings of Saturn. Almost always each ring of vapours ought to be divided into several masses, which, being moved with velocities which differ little from each other, should continue to revolve at the same distance about the Sun. These masses should assume a spheroidical form, with a rotator motion in the direction of that of their revolution, because their inferior particles have a less real velocity than the superior; they have therefore constituted so many planets in a state of vapour. But if one of them was sufficiently powerful, to unite successively by its attraction, all the others about its centre, the ring of vapours would be changed into one sole spheroidical mass, circulating about the Sun, with a motion of rotation in the same direction with that of a revolution. This last case has been the most common; however, the solar system presents to us the first case, in the four small planets which revolve between Mars and Jupiter, at least unless we suppose with Olbers, that they originally formed one planet only, which was divided by an explosion into several parts, and actuated by different velocities. Now if we trace the changes which a farther cooling ought to produce in the planets formed of vapours, and of which we have suggested the formation, we shall see to arise in the centre of each of them, a nucleus increasing continually, by the condensation of the atmosphere which environs it. In this state, the planet resembles the Sun in the nebulous state, in which we have first supposed it to be; the cooling should therefore produce at the different limits of its atmosphere, phenomena similar to those which have been described, namely, rings and satellites circulating about its center in the direction of its motion of rotation, and revolving in the same direction on their axes. The regular distribution of the mass of rings of Saturn about its centre and in the plane of its equator, results naturally from this hypothesis, and, without it, is inexplicable. Those rings appear to me to be existing proofs of the primitive extension of the atmosphere of Saturn, and of its successive condensations. Thus the singular phenomenon of the small eccentricities of the orbits of the planets and satellites, of the small inclination of these orbits to the solar equator, and of the identity in the direction of the motions of rotation and revolution of all those bodies with that of the rotation of the Sun, follow from the hypothesis which has been suggested, and render it extremely probable. If the solar system was formed with perfect regularity, the orbits of the bodies which compose it would be circles, of which the planes, as well as those of the various equators and rings, would coincide with the plane of the solar equator. But we may suppose that the innumerable varieties which must necessarily exist in the temperature and density of different parts of these great masses, ought to produce the eccentricities of their orbits, and the deviations of their motions, from the plane of this equator.

Experiments to Determine the Density of the Earth

by Henry Cavendish (1798)

HENRY CAVENDISH (1731–1810), English chemist and physicist, adopted a strictly experimental approach to the measuring of Earth's mass and density. The key for Cavendish was to determine the gravitational constant, G, in Newton's famous equation for gravity:

$$F_G = G \times [(m_{Earth} \times m_2)/d^2]$$

Cavendish already knew the force (F_G) exerted by an object (of mass, m_2), as well as Earth's radius (d in Newton's equation). Thus, by measuring G, he could calculate Earth's total mass (m_{Earth}), as well as its density.

* * *

Many years ago, the late Rev. JOHN MICHELL, of this Society, contrived a method of determining the density of the earth, by rendering sensible the attraction of small quantities of matter; but, as he was engaged in other pursuits, he did not complete the apparatus till a short time before his death, and did not live to make any experiments with it. After his death, the apparatus came to the Rev. FRANCIS JOHN HYDE WOLLASTON, Jacksonian

Professor at Cambridge, who, not having conveniences for making experiments with it in the manner he could wish, was so good as to give it to me.

The apparatus is very simple; it consists of a wooden arm, 6-feet long, made so as to unite great strength with little weight. This arm is suspended in a horizontal position, by a slender wire 40 inches long, and to each extremity is hung a leaden ball, about 2 inches in diameter; and the whole is inclosed in a narrow wooden case, to defend it from the wind.

As no more force is required to make this arm turn round on its centre, than what is necessary to twist the suspending wire, it is plain, that if the wire is sufficiently slender, the most minute force, such as the attraction of a leaden weight a few inches in diameter, will be sufficient to draw the arm sensibly aside. The weights which Mr. MICHELL intended to use were 8 inches diameter. One of these was to be placed on one side the case, opposite to one of the balls, and as near it as could conveniently be done, and the other on the other side, opposite to the other ball, so that the attraction of both these weights would conspire in drawing the arm aside; and, when its position, as affected by these weights, was ascertained, the weights were to be removed to the other side of the case, so as to draw the arm the contrary way, and the position of the arm was to be again determined; and, consequently, half the difference of these positions would shew how much the arm was drawn aside by the attraction of the weights.

In order to determine from hence the density of the earth, it is necessary to ascertain what force is required to draw the arm aside through a given space. This Mr. MICHELL intended to do, by putting the arm in motion, and observing the time of its vibrations, from which it may easily be computed.

As I was convinced of the necessity of guarding against this source of error [*i.e., changes of temperature*], I resolved to place the apparatus in a room which should remain constantly shut, and to observe the motion of the arm from without, by means of a telescope; and to suspend the leaden weights in such a manner, that I could move them without entering into the room. This difference in the manner of observing, rendered it necessary to make some alteration in Mr. MICHELL'S apparatus; and, as there were some parts of it which I thought not so convenient as could be wished, I chose to make the greatest part of it afresh.

Before I proceed to the account of the experiments, it will be proper to say something of the manner of observing. Suppose the arm to be at rest, and its position to be observed, let the weights then be moved, the arm will not only be drawn aside thereby, but it will be made to vibrate, and its vibrations will continue a great while; so that, in order to determine how much the arm is drawn aside, it is necessary to observe the extreme points of the vibrations, and from thence to determine the point which it would rest at if its motion was destroyed, or the point of rest, as I shall call it. To do this, I observe three successive extreme points of vibration, and take the mean between the first and third of these points, as the extreme point of vibration in one direction, and then assume the mean between this and the second extreme, as the point of rest; for, as the vibrations are continually diminishing, it is evident, that the mean between two extreme points will not give the true point of rest. ...

It appears, therefore, that on account of the resistance of the air, the time at which the arm comes to the middle point of the vibration, is not exactly the mean between the times of its coming to the extreme points, which causes some inaccuracy in my method of finding the time of vibration. It must be observed, however, that as the time of coming to the middle point is before the middle of the vibration, both in the first and last vibration, and in general is nearly equally so, the error produced from this cause must be inconsiderable; and, on the whole, I see no method of finding the time of a vibration which is liable to less objection. ...

In my first experiments, the wire by which the arm was suspended was 39¼ inches long, and was of copper silvered, one foot of which weighed 2 4/10 grains; its stiffness was such, as to make the arm perform a vibration in about 15 minutes. I immediately found, indeed, that it was not stiff enough, as the attraction of the weights drew the balls so much aside, as to make them touch the sides of the case; I, however, chose to make some experiments with it, before I changed it. ...

CONCLUSION

From this table it appears, that though the experiments agree pretty well together, yet the difference between them, both in the quantity of motion of the arm and in the time of vibration, is greater than can proceed merely from the error of observation. As to the difference in the motion of the arm, it may very well be accounted for from the current of air produced by the difference of temperature; but, whether this can account for the difference in the time of vibration, is doubtful. If the current of air was regular and of the same swiftness in all parts of the vibration of the ball, I think it could not; but, as there will most likely be much irregularity in the current, it may very likely be sufficient to account for the difference.

By a mean of the experiments made with the wire first used, the density of the earth comes out 5.48 times greater than that of water; and by a mean of those made with the latter wire, it comes out the same; and the extreme difference of the results of the 23 observations made with this wire, is only .75; so that the extreme results do not differ from the mean by more than .38, or 1/14 of the whole, and therefore the density should seem to be determined hereby, to great exactness. It, indeed, may be objected, that as the result appears to be influenced by the current of air, or some: other cause, the laws of which we are not well acquainted with, this cause may perhaps act always, or commonly, in the same direction, and thereby make a considerable error in the result. But yet, as the experiments were tried in various weathers, and with considerable variety in the difference of temperature of the weights and air, and with the arm resting at different distances from the sides of the case, it seems very unlikely that this cause should act so uniformly in the same way, as to make the error of the mean result nearly equal to the difference between, this and the extreme; and, therefore, it seems very unlikely that the density of the earth should differ from 5.48 by so much as 1/14 of the whole.

According to the experiments made by Dr. MASKELYNE, on the attraction of the hill Schahillien, the density of the earth is 4½ times that of water; which differs rather more from the preceding determination than I should have expected. But I forbear entering into any consideration of which determination is most to be depended on, till I have examined more carefully how much the preceding determination is affected by irregularities whose quantity I cannot measure.

CHAPTER 17
PLATE TECTONICS

*The entire Earth is still changing, due to the slow convection
of soft, hot rocks deep within the planet.*

INTRODUCTION

P RIOR TO THE 1950s, most geologists accepted the concept that Earth's continents
and oceans represent fixed features of the globe and that mountain ranges form
locally, for example by compression of the crust. Two new submarine-hunting
technologies made available to ocean scientists after World War II provided key evidence
that changed geologists' world view and led to the revolutionary theory of plate tectonics.
The following two seminal articles introduced some of that new data.

The first of these technologies was sonar, which uses the reflection of underwater sound
waves to measure distance. Sonar allowed detailed mapping of the topography of the ocean
floor for the first time in the 1950s, when ships made dozens of sonar-tracking voyages across
the Atlantic Ocean. Bruce Heezen (1924–1977) and his Columbia University colleagues
Maria Tharp (1920–2006) and Maurice Ewing (1906–1974) were pioneers in this effort.
These surveys led to the discovery of the Mid-Atlantic Ridge—the longest mountain range
on Earth. In their monograph, *The Floors of the Ocean* (1959), Heezen, Tharp and Ewing
describe the size and shape of the Mid-Atlantic Ridge for the first time, and they speculate
that the immense feature represents a "rift valley."—what we now recognize as a divergent
plate boundary where chains of volcanoes form new crust.

Note that these articles use some unfamiliar oceanographic terms. "Bathymetry" is the
measurement of ocean depths; "fathom" is a unit of water depth measurement of 6 feet; and
an "isobath" is a surface of constant ocean depth.

The Floors of the Ocean
by Bruce C. Heezen, Marie Tharp, and Maurice Ewing ()

PART 2. PHYSIOGRAPHIC PROVINCES
INTRODUCTION

DESCRIPTIONS OF PHYSICAL features of the earth's surface are found in the earliest-known writings. However, the systematic classification of land forms is relatively recent and followed the development of the science of physical geology. The natural topographic divisions of the continents have been classified into physiographic provinces according to several similar systems. These systems take into account form and age of the relief, as well as the structure of the underlying rocks. Descriptions are usually given in terms of age, process, and structure, with the ultimate aim the understanding of the origin and history of topography. Detailed topographic maps at 1:50,000 or larger are available and are used in conjunction with direct field observations. More recently aerial photographs have greatly aided geomorphic studies.

The oceans, in contrast, have been subdivided by oceanographers merely into basins separated by ridges and swells. This was done on the basis of widely spaced discrete soundings shown on charts rarely of larger scale than 1:10 million. The basins were delimited by arbitrarily chosen and often crudely controlled isobaths. The development and installation of continuously recording deep-sea echo sounders and their extensive use in the deep sea provide for the first time detailed topographic information on the deep-sea floor and thus a new basis for description and classification. ...

MID-OCEANIC RIDGE
DEFINITION

The third basic subdivision of the oceanic depression is the Mid-Oceanic Ridge, a continuous median ridge which runs the length of the North Atlantic, South Atlantic, Indian and South Pacific oceans, for more than 40,000 miles. In the center third of the physiographic diagram a short segment of this world-encircling ridge is represented.

MID-ATLANTIC RIDGE

One can find references to the Mid-Atlantic Ridge in the scientific literature dating back more than 80 years. Before the advent of the echo sounder the lateral limits of the Mid-Atlantic Ridge were very difficult to define, and even now widely different definitions are used. Murray (1912) mentioned that the ridge lay in depths less than 2000 fathoms but pointed out that locally on the ridge depths exceeded 2000 fathoms. The METEOR expedition charts and profiles generally imply by their labeling that the ridge is the area enclosed by the 4000-meter contour (2250 fathoms). Shepard (1948) states that its "average depth is about 1500 fathoms, but it rises about 1000 fathoms above deeper zones on either side."

Tolstoy and Ewing (1949) and Tolstoy (1951) in general limit the ridge to depths of less than 2500 fathoms, although in one part of the text Tolstoy and Ewing limit it to less than 2240 fathoms, and Tolstoy (1951) implies that the ridge extends to 2900 fathoms. In the present paper the Mid-Atlantic Ridge is considered as a morpho-tectonic unit defined in terms of morphology, and therefore its definition is not based on a closed isobath.

The Mid-Atlantic Ridge is that portion of the Mid-Oceanic Ridge system which lies within the limits of the Atlantic Ocean. It consists of a broad, fractured median arch or swell which occupies approximately the center third of the ocean. Its crest lies near the median line of the ocean, and its lateral boundaries are formed by scarps which lie near the axes of maximum depth of the eastern and western basins. Adjacent to the Mid-Atlantic Ridge both to the east and to the west is the abyssal floor (usually abyssal hills) of the ocean-basin floor.

PROVINCES OF THE MID-ATLANTIC RIDGE

The Mid-Atlantic Ridge was subdivided by Tolstoy and Ewing (1949) and Tolstoy (1951) into (a) "the central backbone or main range which is shallower than 1600 fathoms," and (b) "the flanks" or "the terraced zone" "between the 1600- and 2500-fathom isobaths." In this paper we use a similar but somewhat differently defined system by dividing the provinces of the Mid-Atlantic Ridge into two categories: (1) the crest provinces, and (2) the flank provinces.

Crest provinces.—The provinces of the crest of the Mid-Atlantic Ridge consist of (1) the Rift Valley (or Valleys); (2) Rift Mountains; and (3) High Fractured Plateau. The Azores Plateau, which forms part of the crest, presents additional problems and is discussed separately.

RIFT VALLEY: The most striking feature on an average profile across the Mid-Atlantic Ridge is a deep notch or cleft in the crest of the ridge. In a small percentage of the sounding profiles two or three such valleys are present, and on a few profiles no notable depressions are observed. On an average profile the floor of the valley lies at about 2000 fathoms, while the adjacent peaks average about 1000 fathoms below the sea surface. The width of the valley between the crests of the adjacent peaks ranges between 15 and 30 miles, and the depth of the valley floor beneath the highest adjacent peak ranges from 700 to 2100 fathoms. The width of the valley 500 fathoms above its floor ranges from 5 to 22 miles. The range in observed depths of the valley is 1150 to 2850 fathoms in the area of the physiographic diagram. The adjacent peaks range from 500 to 1300 fathoms within the same area (excluding the area near the Azores).

Twenty-six crossings of the Rift Valley are shown in Figure 45. The profiles can be divided into three groups: (1) single well-developed rift valley; (2) several well-developed deep valleys; (3) no particularly deep central valley. Most of the profiles (20) fall into the first class; the second class is represented by 5, and only 1 falls in the third class.

The topography of the floor of the rift is rough. In no instance has a flat floor been observed. Where the valley is widest mountains a few hundred fathoms high protrude from its floor.

RIFT MOUNTAINS: The steep walls flanking the rift each form one side of a large rough-sided block. They might be considered as tilted blocks whose facing slopes form the Rift Valley. The back or outer slope of the Rift-Mountains Province is generally broken into mountains as much as 500 fathoms high and 10 miles wide. The lateral limit of the Rift-Mountains Province, is reached when the average slope of the sea floor flattens markedly. Because of the high local relief it is sometimes difficult to pick the boundary of the Rift Mountains, but in almost all recorded profiles the approximate position of the boundary is unmistakable. ...

Origin of the Mid-Atlantic Ridge.—Of the many theories which have been proposed for the origin of the Mid-Atlantic Ridge almost all have been extremely speculative, and none has been based on any very detailed knowledge of the feature. We are still a long way from having a comprehensive knowledge of the Ridge. The various theories of origin and their factual basis have been briefly reviewed by Tolstoy and Ewing, who conclude that it is impossible to say if the feature is primarily of folded or faulted origin. In a paper in press Heezen and Ewing compare in detail the topography and seismicity of the African rift valleys and the Rift Valley of the Mid-Atlantic Ridge. Their conclusion is that the two areas are of basically the same structure, and in fact both form parts of the same continuous structural feature. Since the African rift valleys seem clearly to be the result of normal faulting resulting from extension of the crust, Heezen and Ewing conclude that the topography of the Mid-Atlantic Ridge is largely the result of normal faulting. Whether the forces are the result of horizontal extension or vertical uplift remains the most important unsolved problem in connection with the origin of the continental as well as the sub-oceanic rift-valley systems. Hess (1954) has proposed a mechanism relating suboceanic uplift to expansion due to serpentization of the upper mantle.

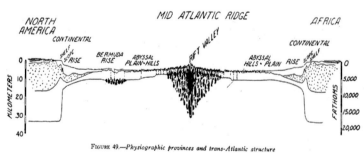

FIGURE 49.—*Physiographic provinces and trans-Atlantic structure*

Magnetic Anomalies Over Ocean Ridges
by Frederick J. Vine and Drummond H. Matthews (1963)

THE "SMOKING GUN" of plate tectonics came from the other new submarine-hunting technology—magnetometry, or the detection of variations in Earth's magnetic field caused by nearby objects made of metal or other magnetic materials. Volcanic rocks on the ocean floor contain the magnetic iron oxide, magnetite, which aligns with Earth's magnetic field as the rock cools from its molten state. The ocean floor thus preserves a record of the direction of Earth's magnetic field. University of Cambridge geophysicists Frederick J. Vine (1939–) and Drummond H. Matthews (1931–1997) were among the pioneers who first measured and recognized the remarkable periodic 180-degree reversal of Earth's magnetic field, as recorded in profiles across the Mid-Atlantic Ridge and other ocean floor ridge systems. These magnetic features occur in long stripes that are parallel to, and symmetric about, the ocean ridges. Vine and Matthews reviewed several possible explanations for their data and arrive at two remarkable conclusions: the orientation of Earth's magnetic field flips, and the floor of the ocean is spreading as new crust is formed at mid-ocean ridges.

* * *

Typical profiles showing bathymetry and the associated total magnetic field anomaly observed on crossing the North Atlantic and North-West Indian Oceans are shown in Fig. 1. They illustrate the essential features of magnetic anomalies over the oceanic ridges: (1) long-period anomalies over the exposed or buried foothills of the ridge; (2) shorter-period anomalies over the rugged flanks of the ridge; (3) a pronounced central anomaly associated with the median valley. This pattern has now been observed in the North Atlantic,[1, 2] the Antarctic,[3] and the Indian Oceans.[4, 5] In this article we describe an attempt to account for it.

The general increase in wave-length of the anomalies away from the crest of the ridge is almost certainly associated with the increase in depth to the magnetic crustal material.[1] Local anomalies of short-period may often be correlated with bathymetry, and explained in terms of reasonable susceptibility contrasts and crustal configurations; but the long-period anomalies of category (1) are not so readily explained. The central anomaly can be reproduced if it is assumed that a block of material very strongly magnetized in the present direction of the Earth's field underlies the median valley and produces a positive susceptibility contrast with the adjacent crust. It is not clear, how over, why this considerable susceptibility contrast should exist beneath the median valley but not elsewhere under the ridge. Recent work in this Department has suggested a new mechanism.

In November 1962, H.M.S. *Owen* made a detailed magnetic survey over a central part of the Carlsberg Ridge as part of the International Indian Ocean Expedition. The area (50 x 40 nautical miles; centred on 5° 25' N., 61° 45' E.) is predominantly mountainous, depths ranging from 900 to 2,200 fathoms, and the topographic features are generally elongated parallel to the trend of the Ridge. This elongation is more marked on the total magnetic field anomaly map where a trough of negative anomalies, flanked by steep gradients, separates two areas of positive anomalies. The trough of negative anomalies corresponds to a general depression in the bottom topography which represents the median valley of the Ridge. The positive anomalies correspond to mountains on either side of the valley.

In this low magnetic latitude (inclination - 6°) the effect of a body magnetized in the present direction of the Earth's field is to reduce the strength of the field above it, producing a negative anomaly over the body and a slight positive anomaly to the north. Here, over the centre of the Ridge, the bottom topography indicates the relief of basic extrusives such as volcanoes and fissure eruptives, and there is little sediment fill. The bathymetry, therefore, defines the upper surface of magnetic material having a considerable intensity of magnetization, potentially as high as any known igneous rock type, and probably higher, because it is extrusive, than the main crustal layer beneath. That the topographic features *are* capable of producing anomalies is immediately apparent on comparing the bathymetric and the anomaly charts; several have well-defined anomalies associated with them.

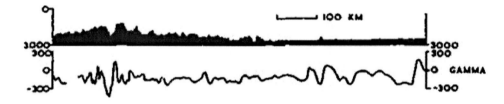

Fig. 1. Profile showing bathymetry and the associated total magnetic field anomaly observed on crossing the northwest Indian Ocean: profile from 30° 5' N. 61° 57' E. to 10° 10' N. 66° 27' E

Two comparatively isolated volcano-like features were singled out and considered in detail. One has an associated negative anomaly as one would expect for normal magnetization, the other, completely the reverse anomaly pattern, that is, a pronounced positive anomaly suggesting reversed magnetization. Data on the topography of each feature and its associated anomaly were fed into a computer and an intensity and direction of magnetization for each obtained. ... Having computed the magnetic vector by a 'best fit' process, the computer recalculated the anomaly over the body, assuming this vector, thus giving an indication of the accuracy of fit. The fit was good for the case of reversed magnetization but poor for that of approximately normal magnetization. The discrepancy is scarcely surprising since we have ignored the effects of adjacent topography, and the interference of other anomalies in the vicinity. In addition, the example of normal magnetization is near a corner of the area where the control of contouring is less precise. The other example is central whore the control is good. In both cases the intensity of magnetization deduced was about 0.005 e.m.u. ...

In addition, three profiles, perpendicular to the trend of the Ridge, have been considered. Computed profiles along these, assuming infinite lateral extent of the bathymetric profile, and uniform normal magnetization, bear little resemblance to the observed profiles. These results suggested that whole blocks of the survey area might be reversely magnetized. ... [We computed] a model in which the main crustal layer and overlying volcanic terrain were divided into blocks about 20 km wide, alternately normally and reversely magnetized. The blocks were given the effective susceptibility values shown in the caption to Fig. 4.

Work on this survey led us to suggest that some 60 per cent of the oceanic crust might be reversely magnetized and this in turn has suggested a new model to account for the pattern of magnetic anomalies over the ridges.

The theory is consistent with, in fact virtually a corollary of, current ideas on ocean floor spreading[7] and periodic reversals in the Earth's magnetic field.[8] If the main crustal layer (seismic layer 3) of the oceanic crust is formed over a convective up-current in the mantle at the centre of an oceanic ridge, it will be magnetized in the current direction of the Earth's field. Assuming impermanence of the ocean floor, the whole of the oceanic crust is comparatively young, probably not older than 150 million years, and the thermo-remanent component of its magnetization is therefore either essentially normal, or reversed with respect to the present field of the Earth. Thus, if spreading of the ocean floor occurs, blocks of alternately normal and reversely magnetized material would drift away from the centre of the ridge and parallel to the crest of it.

This configuration of magnetic material could explain the lineation or 'grain' of magnetic anomalies observed over the Eastern Pacific to the west of North America (probably equivalent to the long-period anomalies of category (1)). Here north-south highs and lows of varying width, usually of the order of 20 km, are bounded by steep gradients. The amplitude and form of these anomalies have been reproduced by Mason,[9, 10] but the most plausible of the models used involved very severe restrictions on the distribution of lava flows in crustal layer 2. They are readily explained in terms of reversals assuming the model shown in Fig. 4 (1). It can be shown that this type of anomaly pattern will be produced for virtually all orientations and magnetic latitudes, the amplitude decreasing as the trend of the ridge approaches north-south or the profile approaches the magnetic equator. The pronounced central anomaly over the ridges is also readily explained in terms of reversals. The central block, being most recent, is the only one which has a uniformly directed magnetic vector. This is comparable to the area of normally magnetized late Quaternary basics in Central Iceland[11, 12] on the line of the Mid-Atlantic Ridge. Adjacent and all other blocks have doubtless been subjected to subsequent vulcanism in the form of volcanoes, fissure eruptions, and lava flows, often oppositely magnetized and hence reducing the effective susceptibility of the block, whether initially normal or reversed. The effect of assuming a reduced effective susceptibility for the adjacent blocks is illustrated for the North Atlantic and Carlsberg Ridges in Fig. 4 (2. 3).

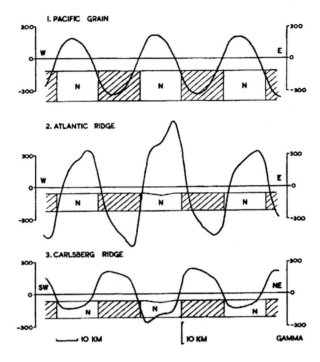

Fig. 4. Magnetic profiles computed for various crustal models. Crustal blocks marked *N*, normally magnetized; diagonally shaded blocks, reversely magnetized. ...

In Fig. 4, no attempt has been made to reproduce observed profiles in detail, the computations simply show that the essential form of the anomalies is readily achieved. The whole of the magnetic material of the oceanic crust is probably of basic igneous composition; however, variations in its intensity of magnetization and in the topography and direction of magnetization of surface extrusives could account for the complexity of the observed profiles. The results from the preliminary Mohole drilling[13, 14] are considered to substantiate this conception. The drill penetrated 40 ft. into a basalt lava flow at the bottom of the hole, and this proved to be reversely magnetized.[13] Since the only reasonable explanation of the magnetic anomalies mapped near the site of the drilling is that the area is underlain by a block of normally magnetized crustal material,[14] it appears that the drill penetrated a layer of reversely magnetized lava overlying a normally magnetized block.

In Fig. 4 it will also be noticed that the effective susceptibilities assumed are two to five times less than that derived for the isolated features in the survey area described. Although no great significance can be attached to this derived intensity it is suggested that the fine-grained extrusives (basalts) of surface features are more highly magnetized than the intrusive material of the main crustal layer which, in the absence of evidence to the contrary, we assume to be of analogous chemical composition (that is, gabbros). This would appear to be consistent with recent investigations of the magnetic properties of basic rocks.[6]

The vertical extent of the magnetic crust is defined by the depth to the curie-point isotherm. To the models this has been assumed to be at 20 km below sea-level over the deep ocean but at a depth of 11 km beneath the centre of the ridges where the heat flow and presumably the thermal gradient are higher. These assumptions are questionable but not critical because the amplitude of the simulated anomaly depends on both the thickness of the block und its effective susceptibility, and. although the thickness is in doubt by a factor of two, the susceptibility is in doubt by a factor of ten. Present magnetic declination has been assumed throughout the calculations: it would probably have been better to have ignored this, us in paleaomagnetism, assuming that true north approximates to the mean of secular variations; but this is unimportant, and in no way affects the essential features of the computations.

In order to explain the steep gradients and largo amplitudes of magnetic anomalies observed over oceanic ridges all authors have been compelled to assume vertical boundaries and high-susceptibility contrasts between adjacent crustal blocks. It is appreciated that magnetic contrasts within the oceanic crust can be explained without postulating reversals of the Earth's magnetic field; for example, the crust might contain blocks of very strongly magnetized material adjacent to blocks of material weakly magnetized in the same direction. However, the model suggested in this article seems to be more plausible because high susceptibility contrasts between adjacent blocks can be explained without recourse to major inhomogeneities of rock type within the main crustal layer or to unusually strongly magnetized rocks.

We thank Dr. R, G. Mason and K. Kunaratnam of the Imperial College of Science and Technology, London, for details of the three-dimensional programme used in this work. The programme was originally devised by K. Kunaratnam for a Ferranti *Mercury* Computer. It has been rewritten for use on *Edsac* 2. We also thank the Director of the Cambridge University Mathematical Laboratory for permission to use *Edsac* 2, and Sir Edward Bullard for his advice and encouragement throughout.

This work was partly supported by a grant from the U.S. Office of Naval Research (Contract No. *N*02558–3542).

1. Heezen, B. C., Ewing, M., and Miller, E. T., *Deep Sea Res.*,1, 25 (1953).
2. Keen, M. J., *Nature*, 197. 88S (1963).
3. Adams, R D., and Christoffel. D. A., *J. Geophys. Res.*, 67. 805 (1962).
4. Heirtzler, J. R., *Tech. Rep.* No.2. *Lamont Geol. Obs., New York* (1961).
5. Matthews, D. H., *el at,. Admiralty Marine Sci. Pub. No.* 4 (in the press).
6. Bullard, E. C., and Mason, R. G., *The Sea.* 3, edit. by Hill, M. N. (in the press).
7. Dietz, R. S., *Nature*, 190. 854 (1961).
8. Cox, A., Doell, R. R., and Dalrymple, G. B., *Nature*, 198, 1049 (1963).

9.   Mason, R. G., *Geophys. J.*, 1, 320 (1058).

10. Mason. R. G., and Raff. A. D., *Bull. Geol. Soc. Amer.*, 72, 1259 (1961).

11. Hospers, J. *Geol. Mag.* 91. 352 (1954).

12. Thorarinsson, S., Einarsson, T., and Kjartansson, G., *Intern. Geog. Cong. (Norden)*. Excursion E.I.1 (1960).

13. Cox. A., and Doell, R. R., *J. Geophys. Res.*, 67. 3997 (1962).

14. Raff. A. D., *J. Geophys. Res.*, 88, 955 (1963).

CHAPTER 18
CYCLES OF THE EARTH

All matter above and beneath the Earth's surface moves in cycles.

INTRODUCTION

SCOTTISH GEOLOGIST JAMES Hutton (1726–1797) is often regarding as the founder of modern geology because of his transformational work, *Theory of the Earth*, published in Edinburgh in 1795. This four-volume treatise incorporates two core ideas, as represented in the two following excerpts.

The first excerpt, taken from the introductory chapter, presents Earth as a kind of machine that operates according to natural laws. Hutton noted that much of Earth's surface is covered with limestone and other fossil-bearing sedimentary rocks that formed on the ocean floor and subsequently have been raised above sea level. Hutton also recognized Earth's inner heat as the energy source for this "engine," and he argued for the revolutionary idea that immense spans of time are required for Earth to undergo dynamic changes.

The second excerpt from Chapter 6 includes Hutton's description of famous rock outcroppings, including sea cliffs at Jedburgh, Scotland. There Hutton observed highly folded rock layers ("schistus strata" or "schisti") that were once flat-lying sediments. The tops of these contorted rocks have been eroded and they are overlain by more layers of flat sediments. This dramatic succession of rocks tells a story of immense age, as successive periods of sedimentation, burial, metamorphism, uplift, and erosion shape Earth's surface.

Hutton's verbose writing style lacked clarity and thus failed to attract many immediate followers. In 1802, Hutton's friend and fellow naturalist John Playfair published a much more readable popular condensation, *Illustrations of the Huttonian Theory of the Earth*, which quickly converted many Earth scientists to the concept of deep time.

Hutton employed a number of now archaic terms. "Indurated" means hardened, as in the process of turning sand into sandstone; "marl" is a fine-grained sediment containing lime; "schistus" refers to highly folded metamorphic rocks that are common in Scotland and formed the lower portions of Hutton's field area.

Theory of the Earth, Volume 1

With Proofs and Illustrations, in Four Parts

by James Hutton
Edinburgh, 1795

CHAPTER I. THEORY OF THE EARTH; OR AN INVESTIGATION OF THE LAWS OBSERVABLE IN THE COMPOSITION, DISSOLUTION, AND RESTORATION, OF LAND UPON THE GLOBE.

WHEN WE TRACE the parts of which this terrestrial system is composed, and when we view the general connection of those several parts, the whole presents a machine of a peculiar construction by which it is adapted to a certain end. We perceive a fabric, erected in wisdom, to obtain a purpose worthy of the power that is apparent in the production of it.

We know little of the earth's internal parts, or of the materials which compose it at any considerable depth below the surface. But upon the surface of this globe, the more inert matter is replenished with plants, and with animal and intellectual beings. ...

To acquire a general or comprehensive view of this mechanism of the globe, by which it is adapted to the purpose of being a habitable world, it is necessary to distinguish three different bodies which compose the whole. These are, a solid body of earth, an aqueous body of sea, and an elastic fluid of air.

It is the proper shape and disposition of these three bodies that form this globe into a habitable world; and it is the manner in which these constituent bodies are adjusted to each other, and the laws of action by which they are maintained in their proper qualities and respective departments, that form the Theory of the machine which we are now to examine.

Let us begin with some general sketch of the particulars now mentioned.

1st, There is a central body in the globe. This body supports those parts which come to be more immediately exposed to our view, or which may be examined by our sense and observation. This first part is commonly supposed to be solid and inert; but such a conclusion is only mere conjecture; and we shall afterwards find occasion, perhaps, to form another judgment in relation to this subject, after we have examined strictly, upon scientific principles, what appears upon the surface, and have formed conclusions concerning that which must have been transacted in some more central part.

2dly, We find a fluid body of water. This, by gravitation, is reduced to a spherical form, and by the centrifugal force of the earth's rotation, is become oblate. The purpose of this fluid body is essential in the constitution of the world; for, besides affording the means of life and motion to a multifarious race of animals, it is the source of growth and circulation to the organized bodies of this earth, in being the receptacle of the rivers, and the fountain of our vapours.

3dly, We have an irregular body of land raised above the level of the ocean. This, no doubt, is the smallest portion of the globe; but it is the part to us by far most interesting. It is upon the surface of this part that plants are made to grow; consequently, it is by virtue of this land that animal life, as well as vegetation, is sustained in this world.

Lastly, We have a surrounding body of atmosphere, which completes the globe. This vital fluid is no less necessary, in the constitution of the world, than are the other parts; for there is hardly an operation upon the surface of the earth, that is not conducted or promoted by its means. It is a necessary condition for the sustenance of fire; it is the breath of life to animals; it is at least an instrument in vegetation; and, while it contributes to give fertility and health to things that grow, it is employed in preventing noxious effects from such as go into corruption. In short, it is the proper means of circulation for the matter of this world, by raising up the water of the ocean, and pouring it forth upon the surface of the earth.

Such is the mechanism of the globe: Let us now mention some of those powers by which motion is produced, and activity procured to the mere machine.

First, There is the progressive force, or moving power, by which this planetary body, if solely actuated, would depart continually from the path which it now pursues, and thus be forever removed from its end, whether as a planetary body, or as a globe sustaining plants and animals, which may be termed a living world. But this moving body is also actuated by gravitation, which inclines it directly to the central body of the sun. Thus it is made to revolve about that luminary, and to preserve its path.

It is also upon the same principles, that each particular part upon the surface of this globe, is alternately exposed to the influence of light and darkness, in the diurnal rotation of the earth, as well as in its annual revolution. In this manner are produced the vicissitudes of night and day, so variable in the different latitudes from the equator to the pole, and so beautifully calculated to equalise the benefits of light, so variously distributed in the different regions of the globe.

Gravitation, and the *vis infita* of matter, thus form the first two powers distinguishable in the operations of our system, and wisely adapted to the purpose for which they are employed.

We next observe the influence of light and heat, of cold and condensation. It is by means of these two powers that the various operations of this living world are more immediately transacted; although the other powers are no less required, in order to produce or modify these great agents in the economy of life, and system of our changing things. ...

There are other actuating powers employed in the operations of this globe, which we are little more than able to enumerate; such are those of electricity, magnetism, and subterraneous heat or mineral fire.

Powers of such magnitude or force, are not to be supposed useless in a machine contrived surely not without wisdom; but they are mentioned here chiefly on account of their general effect; and it is sufficient to have named powers, of which the actual existence is well known, but of which the proper use in the constitution of the world is still obscure. The laws of electricity and magnetism have been well examined by philosophers; but the purposes of those powers in the economy of the globe have not been discovered. Subterraneous fire, again, although the most conspicuous in the operations of this world, and often examined by philosophers, is a power which has been still less understood, whether with regard to its efficient or final cause. It has hitherto appeared more like the accident of natural things, than the inherent property of the mineral region. It is in this last light, however, that I wish to exhibit it, as a great power acting a material part in the operations of the globe, and as an essential part in the constitution of this world.

We have thus surveyed the machine in general, with those moving powers, by which its operations, diversified almost *ad infinitum*, are performed. Let us now confine our view, more particularly, to that part of the machine on which we dwell, that so we may consider the natural consequences of those operations which, being within our view, we are better qualified to examine. …

A solid body of land could not have answered the purpose of a habitable world; for, a soil is necessary to the growth of plants; and a soil is nothing but the materials collected from the destruction of the solid land. Therefore, the surface of this land, inhabited by man, and covered with plants and animals, is made by nature to decay, in dissolving from that hard and, compact state in which it is found below the soil; and this soil is necessarily washed away, by the continual circulation of the water, running from the summits of the mountains towards the general receptacle of that fluid. The heights of our land are thus leveled with the shores; our fertile plains are formed from the ruins of the mountains; and those traveling materials are still pursued by the moving water, and propelled along the inclined surface of the earth. These moveable materials, delivered into the sea, cannot, for a long continuance, rest upon the shore; for, by the agitation of the winds, the tides and currents, every moveable thing is carried farther and farther along the shelving bottom of the sea, towards the unfathomable regions of the ocean.

If the vegetable soil is thus constantly removed from the surface of the land, and if its place is thus to be supplied from the dissolution of the solid earth, as here represented, we may perceive an end to this beautiful machine; an end, arising from no error in its constitution as a world, but from that destructibility of its land which is so necessary in the system of the globe, in the economy of life and vegetation.

The immense time necessarily required for this total destruction of the land, must not be opposed to that view of future events, which is indicated by the surest facts, and most approved principles. Time, which measures everything in our idea, and is often deficient to our schemes, is to nature endless and as nothing; it cannot limit that by which alone it had existence; and, as the natural course of time, which to us seems infinite, cannot be bounded by any operation that may have an end, the progress of things upon this globe, that is, the course of nature, cannot be limited by time, which must proceed in a continual succession. We are, therefore, to consider as inevitable the deduction of our land, so far as effected by those operations which are necessary in the purpose of the globe, considered as a habitable world; and, so far as we have not examined any other part of the economy of nature, in which other operations and a different intention might appear.

We have now considered the globe of this earth as a machine, constructed upon chemical as well as mechanical principles, by which its different parts are all adapted, in form, in quality, and in quantity, to a certain end; an end attained with certainty or success; and an end from which we may perceive wisdom, in contemplating the means employed.

But is this world to be considered thus merely as a machine, to last no longer than its parts retain their present position, their proper forms and qualities? Or may it not be also considered as an organized body? such as has a constitution in which the necessary decay of the machine is naturally repaired, in the exertion of those productive powers by which it had been formed.

This is the view in which we are now to examine the globe; to see if there be, in the constitution of this world, a reproductive operation, by which a ruined constitution may be again repaired, and a duration or stability thus procured to the machine, considered as a world sustaining plants and animals. ...

Now, if we are to take the written history of man for the rule by which we should judge of the time when the species first began, that period would be but little removed from the present state of things. The Mosaic history places this beginning of man at no great distance; and there has not been found, in natural history, any document by which a high antiquity might be attributed to the human race. But this is not the case with regard to the inferior species of animals, particularly those which inhabit the ocean and its shores. We find, in natural history, monuments which prove that those animals had long existed; and we thus procure a measure for the computation of a period of time extremely remote, though far from being precisely ascertained. ...

It is thus that, in finding the relics of sea-animals of every kind in the solid body of our earth, a natural history of those animals is formed, which includes a certain portion of time; and, for the ascertaining this portion of time, we must again have recourse to the regular operations of this world. We shall thus arrive at facts which indicate a period to which no other species of chronology is able to remount.

In what follows, therefore, we are to examine the construction of the present earth, in order to understand the natural operations of time past; to acquire principles, by which we may conclude with regard to the future course of things, or judge of those operations, by which a world, so wisely ordered, goes into decay; and to learn, by what means such a decayed world may be renovated, or the waste of habitable land upon the globe repaired.

CHAPTER VI. THE THEORY OF INTERCHANGING SEA AND LAND, ILLUSTRATED BY AN INVESTIGATION OF THE PRIMARY AND SECONDARY STRATA

… The river Tweed, below Melrose, discovers in its bed the vertical strata of the schistus mountains, and though here these indurated bodies are not veined with quartz as in many places of the mountains, I did not hesitate to consider them as the same species, that is to say, the marly materials indurated and consolidated in those operations by which they had been so much changed in their place and natural position. Afterwards in travelling south, and seeing the horizontal softer strata, I concluded that I had got out of the alpine country, and supposed that no more of the vertical strata were to be observed.

The river Tiviot has made a wide valley as might have been expected, in running over the whole horizontal strata of marly or decaying substances; and the banks of this river declining gradually are covered with gravel and soil, and show little of the solid strata of the country. This, however, is not the case with the Jed, which is to the southward of the Tiviot; that river, in many places, runs upon the horizontal strata, and undermines steep banks, which falling shows high and beautiful sections of the regular horizontal strata. The little rivulets also which fall into the Jed have hollowed out deep gullies in the land, and show the uniformity of the horizontal strata.

In this manner I was disposed to look for nothing more than what I had seen among those mineral bodies, when one day, walking in the beautiful valley above the town of Jedburgh, I was surprised with the appearance of vertical strata in the bed of the river, where I was certain that the banks were composed of horizontal strata. I was soon satisfied with regard to this phenomenon, and rejoiced at my good fortune in stumbling upon an object so interesting to the natural history of the earth, and which I had been long looking for in vain.

Here the vertical strata, similar to those that are in the bed of the Tweed, appear; and above those vertical strata, are placed the horizontal beds, which extend along the whole country.

The question which we would wish to have solved is this; if the vertical strata had been broken and erected under the superincumbent horizontal strata; or if, after the vertical strata had been broken and erected, the horizontal strata had been deposited upon the vertical strata, then forming the bottom of the sea. That strata, which are regular and horizontal in one place, should be found bended, broken, or disordered at another, is not uncommon; it is always found more or less in all our horizontal strata. Now, to what length this disordering operation might have been carried, among strata under others, without disturbing the order and continuity of those above, may perhaps be difficult to determine; but here, in this present case, is the greatest disturbance of the under strata, and a very great regularity among those above. Here at least is the most difficult case of this kind to conceive, if we are to suppose that the upper strata had been deposited before those below had been broken and erected.

Let us now suppose that the under strata had been disordered at the bottom of the sea, before the superincumbent bodies were deposited; it is not to be well conceived, that the vertical strata should in that case appear to be cut off abruptly, and present their regular edges immediately under the uniformly deposited substances above. But, in the case now under consideration, there appears the most uniform section of the vertical strata, their ends go up regularly to the horizontal deposited bodies. Now, in whatever state the vertical strata had been in at the time of this event, we can hardly suppose that they could have been so perfectly cut off, without any relict being left to trace that operation. It is much more probable to suppose, that the sea had washed away the relics of the broken and disordered strata, before those that are now superincumbent had been begun to be deposited. But we cannot suppose two such contrary operations in the same place, as that of carrying away the relics of those broken strata, and the depositing of sand and subtile earth in such a regular order. We are therefore led to conclude, that the bottom of the sea, or surface of those erected strata, had been in very different situations at those two periods, when the relics of the disordered strata had been carried away, and when the new materials had been deposited.

If this shall be admitted as a just view of the subject, it will be fair to suppose, that the disordered strata had been raised more or less above the surface of the ocean; that, by the effects of either rivers, winds, or tides, the surface of the vertical strata had been washed bare; and that this surface had been afterwards sunk below the influence of those destructive operations, and thus placed in a situation proper for the opposite effect, the accumulation of matter prepared and put in motion by the destroying causes. ...

At Siccar Point, we found a beautiful picture of this junction washed bare by the sea. The sand-stone strata are partly washed away, and partly remaining upon the ends of the vertical schistus; and, in many places, points of the schistus strata are seen standing up through among the sand-stone, the greatest part of which is worn away. Behind this again we have a natural section of those sand-stone strata, containing fragments of the schistus.

After this nothing appears but the schistus rocks, until sand-stone and marl again are found at Red-heugh above the vertical strata. From that bay to Fast Castle we had nothing to observe but the schistus, which is continued without interruption to St Abb's Head. Beyond this, indeed, there appears to be something above the schistus; and great blocks of a red whin-stone or basaltes come down from the height and lie upon the shore; but we could not perceive distinctly how the upper mass is connected with the vertical schistus which is continued below.

Our attention was now directed to what we could observe with respect to the schisti, of which we had most beautiful views and most perfect sections. Here are two objects to be held in view, in making those observations; the original formation or stratification of the schisti, and the posterior operations by which the present state of things has been procured. We had remarkable examples for the illustration of both those subjects.

With regard to the first, we have every where among the rocks many surfaces of the erected strata laid bare, in being separated. Here we found the most distinct marks of strata of sand modified by moving water. It is no other than that which we every day observe upon the sands of our own shore, when the sea has ebbed and left them in a waved figure, which cannot be mistaken. Such figures as these are extremely common in our sand-stone strata; but this is an object which I never had distinctly observed in the alpine schisti; although, considering that the original of those schisti was strata of sand, and formed in water, there was no reason to doubt of such a thing being found. But here the examples are so many and so distinct, that it could not fail to give us great satisfaction.

We were no less gratified in our views with respect to the other object, the mineral operations by which soft strata, regularly formed in horizontal planes at the bottom of the sea, had been hardened and displaced. ...

To a person who examines accurately the composition of our mountains, which occupy the south of Scotland, no argument needs be used to persuade him that the bodies in question are not primitive; the thing is evident from inspection, as much as would be the ruins of an ancient city, although there were no record of its history. The visible materials, which compose for the most part the strata of our south alpine schisti, are so distinctly the *debris* and *detritus* of a former earth, and so similar in their nature with those which for the most part compose the strata on all hands acknowledged as secondary, that there can remain no question upon that head. The consolidation, again, of those strata, and the erection of them from their original position, and from the place in which they had been formed, is another question.

But the acknowledging strata, which had been formed in the sea of loose materials, to be consolidated and raised into the place of land, is plainly giving up the idea of primitive mountains. The only question, therefore, which remains to be solved, must respect the order of things, in comparing the alpine schisti with the secondary strata; and this indeed forms a curious subject of investigation.

It is plain that the schisti had been indurated, elevated, broken, and worn by attrition in water, before the secondary strata, which form the most fertile parts of our earth, had existed. It is also certain that the tops of our schistus mountains had been in the bottom of the sea at the time when our secondary strata had begun to be formed; for the pudding-stone on the top of our Lammermuir mountains, as well as the secondary strata upon the vertical schisti of the Alps and German mountains, affords the most irrefragable evidence of that fact.

It is further to be affirmed, that this whole mass of water-formed materials, as well as the basis on which it rested, had been subjected to the mineral operations of the globe, operations by which the loose and incoherent materials are consolidated, and that which was the bottom of the sea made to occupy the station of land, and serve the purpose for which it is destined in the world. This also will appear evident, when it is considered that it has been from the appearances in this very land, independent of those of the alpine schisti, that the present theory has been established.

By thus admitting a primary and secondary in the formation of our land, the present theory will be confirmed in all its parts. For, nothing but those vicissitudes, in which the old is worn and destroyed, and new land formed to supply its place, can explain that order which is to be perceived in all the works of nature; or give us any satisfactory idea with regard to that apparent disorder and confusion, which would disgrace an agent possessed of wisdom and working with design.

CHAPTER 19
ECOLOGY, ECOSYSTEMS,
AND THE ENVIRONMENT

*Ecosystems are interdependent communities of living things that
recycle matter while energy flows through.*

INTRODUCTION

T HAT ALL LIVING things live in communities of interdependent organisms, called
ecosystems, has been recognized for more than a century. However, the quantita-
tive investigation of the distribution of species and the intricacies of food webs
is a much more recent pursuit. The classic work of Princeton University ecologist Robert
MacArthur (1930–1972) and Harvard University biologist Edward O. Wilson (1929–)
on the ecology of animals on islands ("insular zoogeography") brought a sophisticated
mathematical framework to well known qualitative features of islands. For example, the
number of different plant and animal species on an island increases with the area of the
island (illustrated in their figures 1 and 2), while the distance between islands also plays an
important role in the diversity and overlap of populations (their figure 3). Furthermore,
when a barren island is repopulated (for example, following the devastating volcanic explosion in 1883 of the Indonesian island of Krakataua) the rates of species immigration from
other islands and extinction on the repopulated island are closely linked (their figures 4
and 5). The following excerpt presents the less technical introductory and concluding
portions of MacArthur and Wilson's landmark contribution, which was expanded in their
1967 book, *The Theory of Island Biogeography*.

An Equilibrium Theory
of Insular Zoogeography

by Robert H. MacArthur and Edward O. Wilson (1963)

THE FAUNA-AREA CURVE

As THE AREA of sampling A increases in an ecologically uniform area, the number of plant and animal species s increases in an approximately logarithmic manner, or

$$s = bA^k, \qquad (1)$$

where $k < 1$, as shown most recently in the detailed analysis of Preston (1962). The same relationship holds for islands, where, as one of us has noted (Wilson, 1961), the parameters b and k vary among taxa. Thus, in the ponerine ants of Melanesia and the Moluccas, k (which might be called the *faunal coefficient*) is approximately 0.5 where area is measured in square miles; in the Carabidae and herpetofauna of the Greater Antilles and associated islands, 0.3; in the land and freshwater birds of Indonesia, 0.4; and in the islands of the Sahul Shelf (New Guinea and environs), 0.5.

THE DISTANCE EFFECT IN PACIFIC BIRDS

The relation of number of land and freshwater bird species to area is very orderly in the closely grouped Sunda Islands (fig. 1), but somewhat less so in the islands of Melanesia, Micronesia, and Polynesia taken together (fig. 2). The greater variance of the latter group is attributable primarily to one variable, distance between the islands. In particular, the distance effect can be illustrated by taking the distance from the primary faunal "source area" of Melanesia and relating it to faunal number in the following manner. From fig. 2, take the line connecting New Guinea and the nearby Kei Islands as a "saturation curve" (other lines would be adequate but less suitable to the purpose), calculate the predicted range of "satura-tion" values among "saturated" islands of varying area from the curve, then take calculated "percentage saturation" as $s_i \times 100/B_i$, where s_i is the real number of species on any island and B_i the saturation number for islands of that area. As shown in fig. 3, the percentage satura-tion is nicely correlated in an inverse manner with distance from New Guinea. This allows quantification of the rule expressed qualitatively by past authors (see Mayr, 1940) that island faunas become progressively "impoverished" with distance from the nearest land mass.

AN EQUILIBRIUM MODEL

The impoverishment of the species on remote islands is usually explained, if at all, in terms of the length of time species have been able to colonize and their chances of reaching the remote island in that time. According to this explanation, the number of species on islands grows with time and, given enough time, remote islands will have the same number of species as comparable islands nearer to the source of colonization. The following alternative explanation may often be nearer the truth. Fig. 4 shows how the number of new species entering an island may be balanced by the number of species becoming extinct on that island. The descending curve is the rate at which *new* species enter the island by colonization. This rate does indeed fall as the number of species on the islands increases, because the chance that an immigrant be a new species, not already on the island, falls. Furthermore, the curve falls more steeply at first. This is a consequence of the fact that some species are commoner immigrants than others and that these rapid immigrants are likely, on typical islands, to be the first species present. When there are no species on the island ($N = 0$), the height of the curve represents the number of species arriving per unit of time. Thus the intercept, I, is the rate of immigration of species, new or already present, onto the island. The curve falls to zero at the point $N = P$ where all of the immigrating species are already present so that no new ones are arriving. P is thus the number of species in the "species pool" of immigrants. The shape of the rising curve in the same figure, which represents the rate at which species are becoming extinct on the island, can also be determined roughly. In case all of the species are equally likely to die out and this probability is independent of the number of other species present, the number of species becoming extinct in a unit of time is proportional to the number of species present, so that the curve would rise linearly with N. More realistically, some species die out more readily than others and the more species there are, the rarer each is, and hence an increased number of species increases the likelihood of any given species dying out. Under normal conditions both of these corrections would tend to increase the slope of the extinction curve for large values of N. (In the rare situation in which the species which enter most often as immigrants are the ones which die out most readily—presumably because the island is atypical so that species which are common elsewhere cannot survive well—the curve of extinction may have a steeper slope for small N.) If N is the number of species present at the start, then $E(N)/N$ is the fraction dying out, which can also be interpreted crudely as the probability that any given species will die out. Since this fraction cannot exceed 1, the extinction curve cannot rise higher than the straight line of a 45° angle rising from the origin of the coordinates.

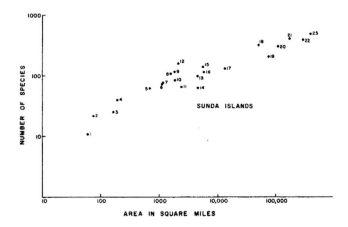

Fig. 1. The numbers of land and freshwater bird species on various islands of the Sunda group, together with the Philippines and New Guinea. The islands are grouped close to one another and to the Asian continent and Greater Sunda group, where most of the species live; and the distance effect is not apparent. (1) Christmas, (2) Bawean, (3) Engano, (4) Savu, (5) Simalur, (6) Alors, (7) Wetar, (8) Nias, (9) Lombok, (10) Buliton, (11) Mentawei, (12) Bali, (13) Sumba, (14) Bangka, (15) Flores, (16) Sumbawa, (17) Timor, (18) Java, (19) Celebes, (20) Philippines, (21) Sumatra, (22) Borneo, (23) New Guinea. Based on data from Delacour and Mayr (1946), Mayr (1940, 1944), Rensch (1936), and Stresemann (1934, 1939).

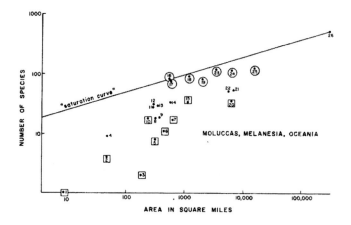

Fig. 2. The numbers of land and freshwater bird species on various islands of the Moluccas, Melanesia, Micronesia, and Polynesia. Here the archiplagoes are widely scattered, and the distance effect is apparent in the greater variance. Hawaii is included even though its fauna is derived mostly from the New World (Mayr, 1943). "Near" islands (less than 500 miles from New Guinea) are enclosed in circles, "far" islands (greater than 2,000 miles) in squares, and islands at intermediate distances are left unenclosed. The saturation curve is drawn through large and small islands at source

of colonization. (1) Wake, (2) Henderson, (3) Line, (4) Kusaie, (5) Tuamotu, (6) Marquesas, (7) Society, (8) Ponape, (9) Marianas, (10) Tonga, (11) Carolines, (12) Patau, (13) Santa Cruz, (14) Rennell, (15) Samoa, (16) Kei, (17) Louisiade, (18) D'Entrecasteaux, (19) Tanimbar, (20) Hawaii, (21) Fiji, (22) New Hebrides, (23) Buru, (24) Ceram, (25) Solomons, (26) New Guinea. Based on data from Mayr (1933, 1940, 1943) and Greenway (1958).

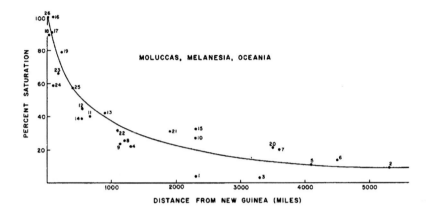

Fig. 3. Per cent saturation, based on the "saturation curve" of fig. 2, as a function of distance from New Guinea. The numbers refer to the same islands identified in the caption of fig. 2. Note that from equation (4) it is an oversimplification to take distances solely from New Guinea. The abscissa should give a more complex function of distances from all the surrounding islands, with the result that far islands would appear less "distant." But this representation expresses the distance effect adequately for the conclusions drawn.

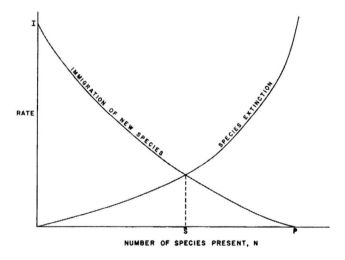

Fig. 4. Equilibrium model of a fauna on a single island. See explanation in text.

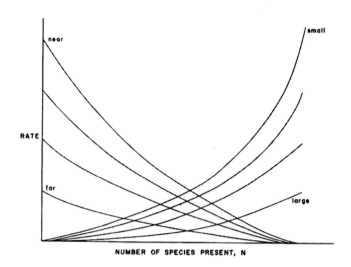

Fig. 5. Equilibrium model of faunas of several islands of varying distances from the source area and varying size. Note that the effect shown by the data of fig. 2, of faunas of far islands increasing with size more rapidly than those of near islands, is predicted by this model. Further explanation in text.

It is clear that the rising and falling curves must intersect and we will denote by s the value of N for which the rate of immigration of new species is balanced by the rate of extinction. The number of species on the island will be stabilized at s, for a glance at the figure shows that when N is greater than s, extinction exceeds immigration of new species so that N decreases, and when N is less than s, immigration of new species exceeds extinction so that N will increase. Therefore, in order to predict the number of species on an island we need only construct these two curves and see where they intersect. We shall make a somewhat over-simplified attempt to do this in later paragraphs. First, however, there are several interesting qualitative predictions which we can make without committing ourselves to any specific shape of the immigration and extinction curves.

A.   An island which is farther from the source of colonization (or for any other reason has a smaller value of I) will, other things being equal, have fewer species, because the immigration curve will be lower and hence intersect the mortality curve farther to the left (see fig. 5).

B.   Reduction of the "species pool" of immigrants, P, will reduce the number of species on the island (for the same reason as in A).

C.   If an island has smaller area, more severe climate (or for any other reason has a greater extinction rate), the mortality curve will rise and the number of species will decrease (see fig. 5).

D.   If we have two islands with the same immigration curve but different extinction curves, any given species on the one with the higher extinction curve is more likely to die out, because $E(N)/N$ can be seen to be higher [$E(N)/N$ is the slope of the line joining the intersection point to the origin].

E. The number of species found on is lands far from the source of colonization will grow more rapidly with island area than will the number on near islands. More precisely, if the area of the island is denoted by A, and s is the equilibrium number of species, then d^2s/dA^2 is greater for far islands than for near ones. This can be verified empirically by plotting points or by noticing that the change in the angle of intersection is greater for far islands.

F. The number of species on large islands decreases with distance from source of colonization faster than does the number of species on small islands. (This is merely another way of writing E and is verified similarly.)

Further, as will be shown later, the variance in s (due to randomness in immigrations and extinctions) will be lower than that expected if the "classical" explanation holds. In the classical explanation most of those species will be found which have at any time succeeded in immigrating. At least for distant islands this number would have an approximately Poisson distribution so that the variance would be approximately equal to the mean. Our model predicts a reduced variance, so that if the observed variance is significantly smaller than the mean for distant islands, it is evidence for the equilibrium explanation.

The evidence in fig. 2, relating to the insular bird faunas east of Weber's Line, is consistent with all of these predictions. To see this for the non-obvious prediction E, notice that a greater slope on this log-log plot corresponds to a greater second derivative, since A becomes sufficiently large. ...

SUMMARY

A graphical equilibrium model, balancing immigration and extinction rates of species, has been developed which appears fully consistent with the fauna-area curves and the distance effect seen in land and freshwater bird faunas of the Indo-Australian islands. The establishment of the equilibrium condition allows the development of a more precise zoogeographic theory than hitherto possible.

One new and non-obvious prediction can be made from the model which is immediately verifiable from existing data, that the number of species increases with area more rapidly on far islands than on near ones. Similarly, the number of species on large islands decreases with distance faster than does the number of species on small islands.

As groups of islands pass from the unsaturated to saturated conditions, the variance-to-mean ratio should change from unity to about one-half. When the faunal buildup reaches 90% of the equilibrium number, the extinction rate in species/year should equal 2.303 times the variance divided by the time (in years) required to reach the 90% level. The implications of this relation are discussed with reference to the Krakatau faunas, where the buildup rate is known.

A "radiation zone," in which the rate of intra-archipelagic exchange of autochthonous species approaches or exceeds extra-archipelagic immigration toward the outer limits of the taxon's range, is predicted as still another consequence of the equilibrium condition. This condition seems to be fulfilled by conventional information but cannot be rigorously tested with the existing data.

Where faunas are at or near equilibrium, it should be possible to devise indirect estimates of the actual immigration and extinction rates, as well as of the times required to reach equilibrium. It should also be possible to estimate the mean dispersal distance of propagules overseas from the zoogeographic data. Mathematical models have been constructed to these ends and certain applications suggested.

The main purpose of the paper is to express the criteria and implications of the equilibrium condition, without extending them for the present beyond the Indo-Australian bird faunas.

CHAPTER 20
STRATEGIES OF LIFE

Living things use many strategies to deal with the problems of acquiring and using matter and energy.

INTRODUCTION

T HE FIRST STEP in any scientific endeavor is to catalog objects and to describe their properties. The following two excerpts appeared in the transformational century following the discovery of Newton's laws of motion and gravity, at a time when scientists were first attempting a modern systematic description of living organisms.

To scientists of the seventeenth century, the mechanisms of plant growth were a mystery. A common assumption held that the solid material of plant tissue derived from soil, which must therefore be consumed as plants grow. Flemish physician Jan Baptist Van Helmont (1577–1644) conducted experiments on plant growth as part of his investigations of the "true elements," which included air and water. He found that the quantity of soil change little as plants grow, whereas copious amounts of water are consumed. He concluded that plants arise primarily from water. Van Helmont's discovery was a major advance, though we now recognize that both water and the gas carbon dioxide in the air are the principal ingredients for photosynthetic organisms. (Ironically, it was Van Helmont, himself, who discovered carbon dioxide, but did not realize its central role in plant growth.)

197

Oriatricke, or Physick Refined, the Common Errors Therein Refuted, and the Whole Art Reformed and Rectified[1]

by John Baptista Van Helmont
Amsterdam, 1648

I HAVE SAID, that there are two primary Elements; the Air, and the Water; because they do not turn into each other: but, that the Earth is as it were born of water; because it may be reduced to water. But if water be changed into an Earthy Body, that happens by the force or virtue of the Seed, and so it hath then put off the simpleness of an Element. For a flint is of water, which is broken asunder into Sand. But surely, that Sand doth lesse resist in its reducing into water, than the Sand, which is the Virgin-Earth. Therefore the Sand of Marble, of a Gemme, or Flint, do disclose the presence of the Seed. But if the Virgin-Earth, may at length, by much labour be brought into water, and if it was in the beginning created as an Element; yet it seemes then to have come down to something that is more simple than it selfe; and therefore I have called those two, Primary ones. I have denied the fire to be an Element and Substance; but to be death in the hand of the Artificer, given for great uses. I say, an artificial Death for Arts, which the Almighty hath created; but not a natural one. ...

But I have learned by this handicraft-operation, that all Vegetables do immediately, and materially proceed out of the Element of water onely. For I took an Earthen Vessel, in which I put 200 pounds of Earth that had been dried in a Furnace, which I moystened with Rain-water, and I implanted therein the Trunk or Stem of a Willow Tree, weighing five pounds; and at length, five years being finished, the Tree sprung from thence, did weigh 169 pounds, and about three ounces: But I moystened the Earthen Vessel with Rain-water or distilled water (always when there was need) and it was large, and implanted into the Earth, and least the dust that flew about should be co-mingled with the Earth, I covered the lip or mouth of the Vessel, with an Iron-Plate covered with Tin, and easily passable with many holes. I computed not the weight of the leaves that fell off in the four Autumnes. At length, I again dried the Earth of the Vessel, and there were found the same 200 pounds, wanting about two ounces. Therefore 164 pounds of Wood, Barks, and Roots, arose out of water onely.

---

1   Translated into English by John Chandler, London, 1662.

The Families of Plants

With their Natural Characters, According to the Number, Figure, Situation, and Proportion of all the Parts of Fructification[1]

by Carolus Linnaeus
Leyden, The Netherlands, 1737

THE CLASSIFICATION OF the overwhelming diversity of plants and animals became the life's work of Swedish naturalist Carl van Linné (1707–1778), whose name is usually Latinized to Carolus Linnaeus. His most lasting contribution is the development of the standard binomial nomenclature, in which each living species is characterized in a hierarchical scheme, starting with the plant and animal kingdoms, and with progressively more restrictive categories: phylum, class, order, family, genus, and species. Thus, organisms are classified systematically according to their similarities and differences. The following excerpt from Linnaeus's 1737 treatise, *Genera Plantarum*, outlines his classification strategy and defines more than two dozen characteristics that he used to distinguish among plant genera.

* * *

ACCOUNT OF THE WORK.

1.   All the real knowledge, which we possess, depends on METHOD; by which we distinguish the similar from the dissimilar. The greater number of natural distinctions this method comprehends, the clearer becomes our idea of the things. The more numerous the objects, which employ our attention, the more difficult it becomes to form such a method; and the more necessary. The great Creator has in no part of his works presented a greater variety to the human mind than in the vegetable kingdom; which covers the whole globe, which we inhabit; whence, if a distinct method is ever necessary, it is necessary here; if we hope to gain a distinct knowledge of vegetables …

2.   To him therefore Vegetables are known, who can join the similar to the similar, and can separate the dissimilar from the dissimilar.

3.   The BOTANIST is he, who can affix similar names to similar vegetables, and different names to different ones, so as to be intelligible to every one.

---

1   Translated from the … *Genera Plantarum* by a BOTANICAL SOCIETY at Lichfield, 1787.

4.   The NAMES of Plants are *generic*, and (where there are any species), *specific*. These should be certain and well founded, not vague, evasive, or variously applicable. Before they can be such, it is necessary, that they should have been affix'd to certain, not to vague genera for if this foundation be unsteady, the names also, and in consequence the doctrine of the Botanist crumbles into ruin.

5.   The SPECIES are as numerous as the different and constant forms of vegetables, which exist upon this globe; which forms according to instinctive laws of generation produce others, similar to themselves, but in greater numbers. Hence there are as many Species, as there are different forms or structures of Plants now existing; excepting such less-different *Varieties*, which situation or accident has occasion'd.

6.   The GENERA are as numerous, as the common proximate attributes of the different species, as they were created in the beginning; this is confirm'd by revelation, discovery, observation, hence

THE GENERA ARE ALL NATURAL

For we must not join in the same genus the horse and the swine, tho' both species had been one-hoof 'd, nor separate in different genera the goat, the raindeer, and the elk, tho' they differ in the form of their horns. We ought therefore by attentive and diligent observation to determine the limits of the genera, since they cannot be determin'd a priori. This is the great work, the important labour. ...

7.   That it has pleased Infinite Wisdom to distinguish the Genera of Plants by their FRUCTIFICATION was discover'd in the last age; and first indeed by CONRADUS GESNER, the ornament of his time; as appears from his posthumous epistles, and from the plates published by CAMERARIUS, altho' the first, who introduced this great discovery into use, was ANDREAS CESALPINUS; which would nevertheless have shortly expired in its cradle, unless it had been recalled into life by the care of ROBERT MORISON, and nourish'd by JOSEPH P. TOURNEFORT with pure systematic rules. This was at length confirm'd by all the great men, then existing, in the science.

8.   This foundation being given, this point fix'd, immediately every one capable of such researches join'd their labours to turn it into use, to build a system; all with the same inclination, and to the same purpose, but not all with equal success. Because the fundamental rule was known but to few, which if the builders did not observe, quickly fell to the ground with the first tempest the insubstantial edifice, however splendid; Boerhave well observes, that "*a* TEACHER, *as he explains discoveries, may proceed from generals to particulars, but an* INVENTOR *on the contrary must pass from particulars to generals.*"

For some assuming the different parts of Fructification as the principle of their system, and descending according to the laws of division from classes through orders even to species, broke and dilacerated the natural Genera, and did violence to nature by their hypothetical and arbitrary principles. For example, one from the *Fruit* denies that the Peach and Almond can be join'd in the same genus, another from the *regularity of the Petals* denies that Valerian and Valerianoides; another from their *number* that Flax and Radiola; another from their *sex* that the urtica androgyna, and dioica, &c. can be combined under the same genus: for, say they, if these cannot be conjoin'd in the same class, much less in the same genus: not having observed, that themselves have contrived the classes, but that the Creator himself made the Genera. Hence arose so many false Genera! such controversy amongst authors! so many bad names! such confusion! Such indeed was the state of things, that as often as a new System-maker arose, the whole Botanic world was thrown into a panic. And for my part I know not, whether these System-builders produced more evil than good to the science; this is certain, if the unlearned be compared with the learned, they much surpass them in number, Physicians, Apothecaries, Gardiners lamented this misfortune, and with reason. I confess, their theory had been excellent, had it pleased the great Creator to have made all the fructifications of the same Genus of Plant equally similar amongst themselves as are the individuals of the same species. As this is not so, we have no resource, since we are not the governors of Nature, nor can create plants according to our own conceptions, but to submit ourselves to the laws of nature, and learn by diligent study to read the characters inscribed on plants. If every different mark of the fructification be adjudged sufficient for distinguishing the genus, why should we a moment hesitate to proclaim nearly as many genera as species? for we are scarcely acquainted with any two flowers so similar to each other, but that some difference of their parts may be discern'd; I also once endeavour'd from the flower alone to determine all the specific differences, but frequently with less success, since there is an easier method. I wish it therefore to be acknowledged by all true Botanists, if they ever expect any certainty in the science, that *the Genera and Species must be all natural* without which assumed principle there can be nothing excellent done in the science. ...

9. I do not deny, that natural Classes may not be given as well as natural Genera. I do not deny, that a natural method ought much to be prefer'd to ours, or those of other discoverers? but I laugh at all the natural methods hitherto cry'd up; and provok'd in my own defence I venture to affirm, that not a single class before given, in any system, is natural so long as those Genera, and those Characters, which at present exist, are arranged under it. It is easy to distribute the greatest part of the known genera into their natural classes, but so much the greater difficulty attends the arrangement of the rest. And I can not persuade myself that the present age will see a natural System, nor perhaps our latest posterity. Let us neverthe-less study plants, and in the mean time content ourselves with artificial and succedaneous Classes. ...

10. These natural Genera assumed, two things are required to preserve them pure, first that the true species, and no others, be reduced to their proper genera. Secondly that all the Genera be circumscribed by true limits or boundaries, which we term *generic characters*.

11. These *characters* as I turn over the authors, I find uncertain and unfix'd before *Tournefort*, to him therefore we ought deservedly to ascribe the honor of this discovery of ascertaining the genera, indeed other Systematists of other Sects have given characters, but I can understand none of them, who proceeded *Tournefort*, or who did not tread in his steps. … *Tournefort* assumed the petals and the fruit, as diagnostic marks of the genera, and no other parts; so did almost all his followers; but the moderns, oppress'd by the quantity of new and lately detected Genera, have supposed that these parts alone are insufficient for distinguishing all the Genera: and have thence believed themselves necessitated to have recourse to the habit and appearance of plants, as the leaves, situation of the flowers, stem, root, &c. that is to recede from the steady foundation of the fructification, and to relaps into the former barbarism; with what ill omen this is done, it would be easy to demonstrate, if this were the time and place for such a talk; whatever may be the event, I acknowledge the parts described by *Tournefort* to be insufficient, if the petals alone or the fruit were to be used for this purpose. But I ask for what reason should these characters alone be used? does inspection shew this? or revelation? or any arguments either a priori, or posteriori ? certainly none of these. I acknowledge no authority but inspection alone in Botany; are there not many more parts of the fructification? why should those only be acknowledged and no others? Did not the same Creator make the latter, as well as the former? are not all the parts equally necessary to the fructification? We have described of the CALYX: I, the *Involucre*; 2, the *Spathe*; 3, the *Perianth*; 4, the *Ament*; 5, the *Glume*; 6, the *Calyptre* of the COROL; 7, the *Tube* or Claws; 8, the *Border*; 9, the *Nectary* of the STAMENS; 10, the *Filaments*; 11, the *Anthers* of the PISTIL; 12, the *Germ*; 13, the *Style*; 14, the *Stigma* of the PERICARP; 15, the *Capsule*; 16, the *Silique*; 17, the *Legume*; 18, the *Nut*; 19, the *Drupe*; 20, the *Berry*; 21, the *Pome* the SEED; 22, 23, and its *crown* the RECEPTACLE; 24, of the *fructification*; 25, of the *flower*; 26, of the *fruit*. Thus are there more parts, more letters here, than in the alphabets of languages. These marks are to us as so many vegetable letters; which, if we can read, will teach us the characters of plants: they are written by the hand of God; it should be our study to read them.

12. *Tournefort* did wonders with his characters, but since so many and such new genera have been since discovered, it should be our business to adhere indeed to his principles, but to augment them with new discoveries, as the science increases.

13. *Figures* alone for determining the genera I do not recommend; before the use of letters was known to mankind, it was necessary to express everything by picture, where it could not be done by word of mouth, but on the discovery of letters the more easy and certain way of communicating ideas by writing succeeded. …

a. From a figure alone who could ever argue with any certainty? but most easily from written words.

b.   If I wish to bring into use, or quote in any work the characters of a Genus; I can not always easily paint the figure of it, nor etch or ingrave it, or print it off, but can easily copy the description.

c.   If in the same Genus, as happens in many, some of the parts should differ in respect to number or figure in some of the species or individuals, yet I am expected to note the situation and proportion of these parts; it becomes impossible to express these by a print, unless I should give a number of figures. Hence if there should be fifty species, I must exhibit as many prints, and who would be able to extract any certainty from such a variety? But in a description the parts are omitted, which differ; and the labour is much less to describe, those which agree, and much more easy to be understood.

14.   We have therefore endeavour'd to express by words all the marks or distinctions as clearly, if not more so, than others have done by their expensive prints. …

19.   I have selected the *marks* in describing every part of the fructification, which are certain and real, not those which are vague and fluctuating. Others have frequently assumed the taste, the odour, the size (without the proportion); these you will never find adduced by me, but only these four certain immutable mechanic principles: *Number, Figure, Situation,* and *Proportion.* These four attributes, with those twenty-six letters, above-mentioned, distinguish the genera so certainly from each other, that nothing more is wanted. There are other marks for distinguishing genera, but these alone being consider'd, the rest become superfluous, nor is there any necessity to fly to the habit of the plant

22.   I am confident the *Flower* is much to be prefer'd to the *Fruit* in determining the genera, though others have been of a quite different opinion; and that the *Nectaries* are of greater advantage in determining the Genera, than almost any other part; altho' so much neglected and overlook'd by others, that they even had not a name given to them. …

29.   The use of some Botanic System I need not recommend even to beginners, since without system there can be no certainty in Botany. Let two enquirers, one a Systematic, and the other an Empiric, enter a garden fill'd with exotic and unknown plants, and at the same time furnish'd with the best Botanic Library; the former will easily reduce the plants by studying the letters inscribed on the fructification, to their Class, Order, and Genus; after which there remains but to distinguish a few species. The latter will be necessitated to turn over all the books to read all the descriptions, to inspect all the figures with infinite labour; nor unless by accident can be certain of his plant.

CHAPTER 21
THE LIVING CELL

Life is based on chemistry and chemistry takes place in the cell.

INTRODUCTION

T HE INVENTION OF the microscope was to biology what the telescope was to astronomy—a window into a previously unseen world. The following two excerpts, written more than two centuries apart, reflect the importance of the microscope in discovering the characteristics of life at scales too small to see with the unaided eye.

The microscope was the co-invention principally of the Dutch lens maker Anthony van Leeuwenhoek (1632–1723) and British physicist and inventor Robert Hooke (1635–1703). Both men made pioneering observations with first-generation microscopes of their own design, and both discovered myriad life forms too small to be seen with the unaided eye. Hooke's classic treatise *Micrographia* (London, 1665) included the first use of the word "cell," and illustrated a variety of organisms found in the local water supply. In this except Hooke describes the "water insect or gnat," which has a larval stage in water, and then metamorphoses into a flying insect.

Micrographia

Or Some Physiological Descriptions of Minute Bodies
Made by Magnifying Glasses. With Observations
and Enquiries Thereupon.

by Robert Hooke
London, 1665

OBSERV. XLIII. OF THE WATER-INSECT OR GNAT

THIS LITTLE CREATURE, described in the first *Figure* of the 27. *Scheme*, was a small scaled or crusted Animal which I have often observ'd to be generated in Rain-water; I have also observ'd it both in Pond and River-water. It is suppos'd by some, to deduce its first original from the putrifaction of Rain-water, in which, if it have stood any time open to the air, you shall seldom miss, all the Summer long, of store of them striking too and fro.

'Tis a creature, wholly differing in shape from any I ever observed, nor is its motion less strange: It has a very large head, in proportion to its body, all covered with a shell, like other *testaceous* Animals, but it differs in this, that it has, up and down several parts of it, several tufts of hairs, or bristles, plac'd in the order express'd in the Figure. It has two horns, which seem'd almost like the horns of an Oxe, inverted, and, as neer as I could guess, were hollow, with tufts of brisles, likewise at the top: these horns they could move easily this or that way, and might perchance, be their nostrils. It has a pretty large mouth, which seem'd contriv'd much like those of Crabs and Lobsters, by which, I have often observ'd them to feed on water, or some imperceptible nutritive substance in it.

I could perceive, through the transparent shell, while the Animal surviv'd, several motions in the head, thorax, and belly, very distinctly, of differing kinds which I may, perhaps, elsewhere endeavour more accurately to examine, and to shew of how great benefit the use of a *Microscope* may be for the discovery of Nature's course in the operations perform'd in Animal bodies, by which we have the opportunity of observing her through these delicate and pellucid teguments of the bodies of Insects acting according to her usual course and way, undisturbed, whereas, when we endeavour to pry into her secrets by breaking open the doors upon her, and dissecting and mangling creatures while there is life yet within them, we find her indeed at work, but put into such disorder by the violence offer'd, as it may easily be imagin'd, how differing a thing we should find, if we could, as we can with a *Microscope*, in these smaller creatures, quietly peep in at the windows, without frighting her out of her usual byas.

The form of the whole creature, as it appear'd in the *Microscope* may, without troubling you with more descriptions, be plainly enough perceiv'd by the *scheme*, the hinder part or belly consisting of eight several jointed parts, namely, *ABCDEFGH*, of the first *Figure*, from the midst of each of which, on either side, issued out three or four small bristles or hairs, *I, I, I, I, I,* the tail was divided into two parts of very differing make; one of them, namely, *Ky* having many tufts of hair or bristles, which seem'd to serve both for the finns and tail, for the Oars and Ruder of this little creature, wherewith it was able, by striking and bending its body nimbly to and fro, to move himself any whither, and to skull and steer himself as he pleas'd; the other part, *L*, seem'd to be, as 'twere, the ninth division of his belly, and had many single bristles on either side. From the end *V*, of which, through the whole belly, there was a kind of Gut of a darker colour, *MMM*, wherein, by certain *Peristaltick* motions there was a kind of black substance mov'd upwards and downwards through it from the orbicular part of it, *N*, (which seem'd the *Ventricle*, or stomach) to the tail *V*, and so back again, which *peristaltick* motion I have observ'd also in a Louse, a Gnat, and several other kinds of transparent body'd Flies. The *Thorax* or chest of this creature *OOOO*, was thick and short,

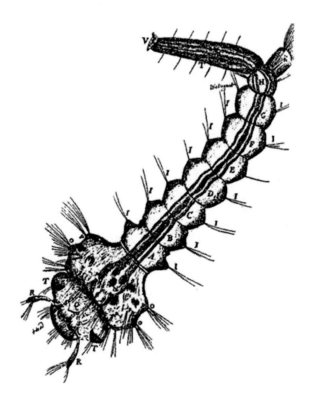

Figure 1.

and pretty transparent, for through it I could see the white heart (which is the colour also of the blood in these, and most other Insects) to beat, and several other kind of motions. It was bestuck and adorn'd up and down with several tufts of bristles, such as are pointed out by *P, P, P, P,* the head *Q* was likewise bestuck with several of those tufts, *SSS*; it was broad and short, had two black eyes, *TT,* which I could not perceive at all pearl'd, as they afterwards appear'd, and two small horns, *RR,* such as I formerly describ'd.

Both its motion and rest is very strange, and pleasant, and differing from those of most other creatures I have observ'd; for, where it ceases from moving its body, the tail of it seeming much lighter than the rest of its body, and a little lighter than the water it swims in, presently boys it up to the top of the water, where it hangs suspended with the head always downward; and like our *Antipodes,* if they do by a frisk get below that superficies, they presently ascend again into it, if they cease moving, until they tread, as it were, under that superficies with their tails; the hanging of these in this posture, put me in mind of a certain creature I have seen in *London,* that was brought out of *America,* which would very firmly suspend it self by the tail, with the head downwards, and was said to sleep in that posture, with her young ones in her false belly, which is a Purse, provided by Nature for the production, nutrition, and preservation of the young ones, which is describ'd by *Piso* in the 24. Chapter of the Fifth Book of his Natural History of *Brazil.*

The motion of it was with the tail forwards, drawing its self backwards, by the striking to and fro of that tuft which grew out of one of the stumps of its tail. It had another motion, which was more sutable to that of other creatures, and that is, with the head forward; for by the moving of his chaps (if I may so call the parts of his mouth) it was able to move itself downwards very gently towards the bottom, and did, as 'twere, eat up its way through the water.

But that which was most observable in this creature, was, its Metamorphosis or change; for having kept several of these Animals in a Glass of Rain-water, in which they were produc'd, I found, after about a fortnight or three weeks keeping, that several of them flew away in Gnats, leaving their hulks behind them in the water floating under the surface, the place where these Animals were wont to reside, whilst they were inhabitants of the water; this made me more diligently to watch them, to see if I could find them at the time of their transformation; and not long after, I observed several of them to be changed into an unusual shape, wholly differing from that they were of before, their head and body being grown much bigger and deeper, but not broader, and their belly, or hinder part smaller, and coyl'd, about this great body much of the fashion represented by the prick'd line in the second *Figure* of the 27 *Scheme,* the head and horns now swam uppermost, and the whole bulk of the body seem'd to be grown much lighter; for when by my frighting of it, it would by striking out of its tail… sink itself below the surface towards the bottom; the body would more swiftly reascend, than when it was in its former shape.

I still marked its progress from time to time, and found its body still to grow bigger and bigger, Nature, as it were, fitting and accoutring it for the lighter Element, of which it was now going to be an inhabitant; for, by observing one of these with my *Microscope*, I found the eyes of it to be altogether differing from what they seem'd before, appearing now all over pearl'd or knobb'd, like the eyes of Gnats, as is visible in the second *Figure* by A. At length, I saw part of this creature to swim above, and part beneath the surface of the water, below which though it would quickly plunge itself if I by any means frighted it, and presently re-ascend into its former posture; after a little longer expectation, I found that the head and body of a Gnat, began to appear and stand clear above the surface, and by degrees it drew out its leggs, first the two formost, then the other, at length its whole body perfect and entire appear'd out of the husk (which it left in the water) standing on its leggs upon the top of the water, and by degrees it began to move, and after flew about the Glass a perfect Gnat.

Figure 2.

I have been the more particular, and large in the relation of the transformation of divers of these little Animals which I observ'd, because I have not found that any Authour has observ'd the like; and because the thing itself is so strange and heterogeneous from the usual progress of other Animals, that I judge it may not only be pleasant, but very usefull and necessary towards the completing of Natural History...

But to return to the more immediate consideration of our Gnat: We have in it an instance, not usual or common, of a very strange *amphibious* creature, that being a creature that inhabits the Air, does yet produce a creature, that for some time lives in the water as a Fish, though afterward (which is as strange) it becomes an inhabitant of the Air, like its Sire, in the form of a Fly. And this, me thinks, does prompt me to propose certain conjectures, as Queries, having not yet had sufficient opportunity and leisure to answer them my self from my own Experiments or Observations.

And the first is, whether all those things that we suppose to be bred from corruption and putrifaction, may not be rationally suppos'd to have their origination as natural as these Gnats, who, 'tis very probable, were first dropt into this Water, in the form of Eggs. Those Seeds or Eggs must certainly be very small, which so small a creature as a Gnat yields, and therefore we need not wonder that we find not the Eggs themselves, some of the younger of them, which I have observ'd, having not ex-seeded a tenth part of the bulk they have afterwards come to; and next, I have observed some of those little ones which must have been generated after the Water was inclosed in the Bottle, and therefore most probably from Eggs, whereas those creatures have been suppos'd to be bred of the corruption of the Water, there being not formerly known any probable way how they should be generated.

A second is, whether these Eggs are immediately dropt into the Water by the Gnats themselves, or, immediately, are brought down by the falling rain; for it seems not very improbable, but that these small seeds of Gnats may (being, perhaps, of so light a nature, and having so great a proportion of surface to so small a bulk of body) be ejected into the Air, and so, perhaps, carried for a good while too and fro in it, till by the drops of Rain it be wash'd out of it.

A third is, whether multitudes of those other little creatures that are found to inhabit the Water for some time, do not, at certain times, take wing and fly into the Air, others dive and hide themselves in the Earth, and so contribute to the increase both of the one and the other Element.

Cellular Pathology
As Based Upon Physiological and Pathological Histology[1]
by Rudolf Virchow (1860)

THE CELL THEORY was first proposed in 1838 by the German botanist Matthais Schleiden (1804–1881), who proposed that all plants are made of cells. In the following year his countryman, zoologist Theodor Schwann (1810–1882), extended this idea to animals and proposed what is now known as the cell theory. Three tenets of the Schwann's cell theory are:

1. All living things are composed of cells.
2. The cell is the fundamental unit of life.
3. All cells arise from previous cells.

The second excerpt, which outlines the cell theory and its importance to medicine, is from *Cellular Pathology* by Rudolf Virchow (1821–1902). Virchow reviews the cell theory and points out similarities and differences between plant and animal cells.

* * *

---

1   German translation by Frank Chance.

LECTURE I
FEBRUARY 10, 1858

CELLS AND THE CELLULAR THEORY

GENTLEMEN,—Whilst bidding you heartily welcome to which must have long since ceased to be familiar to you, I must begin by reminding yon, that it is not my want of modesty which has summoned you hither, but that I have only yielded to the repeatedly manifested wishes of many among you. Nor should I have ventured either to offer you lectures after the same fashion in which I am accustomed to deliver them in my regular courses. On the contrary, I will make the attempt to lay before you in a more succinct manner the development which I myself, think, medical science also, have passed through, of the last fifteen years. In my announcement of these lectures, I described the subject of them in such a way as to couple histology with pathology; and for this reason, that I thought I must take it for granted that many busily occupied physicians were not quite familiar with the most recent histological changes, and did not enjoy sufficiently frequent opportunities of examining microscopical objects for themselves. Inasmuch as, however, it is upon such examinations that the most important conclusions are grounded which we now draw, you will pardon me if, disregarding those among you who have a perfect acquaintance with the subject, I behave just as if you all were not completely familiar with the requisite preliminary knowledge.

The present reform in medicine, of which you have all been witnesses, essentially had its rise in new anatomical obscurations, and the exposition also, which I have to make to you, will therefore principally be based upon anatomical demonstrations. But for me it would not be sufficient to take, as has been the custom during the last ten years, pathological anatomy alone as the groundwork of my views; we must add thereto those facts of general anatomy also, to which the actual state of medical science is due. The history of medicine teaches us, if we will only take a somewhat comprehensive survey of it, that at all times permanent advances have been marked by anatomical innovations, and that every more important epoch has been directly ushered in by a series of important discoveries concerning the structure of the body. So it was in those old times, when the observations of the Alexandrian school, based for the first time upon the anatomy of man, prepared the way for the system of Galen; so it was, too, in the Middle Ages, when Vesalius laid the foundations of anatomy, and therewith began the real reformation of medicine; so, lastly, was it at the commencement of this century, when Bichat developed the principles of general anatomy. What Schwann, however, has done for histology, has as yet been but in a very slight degree built tip and developed for pathology, and it may be said that nothing has penetrated less deeply into the minds of all than the cell-theory in its intimate connection with pathology.

If we consider the extraordinary influence which Bichat in his time exercised upon the state of medical opinion, it is indeed astonishing that such a relatively long period should have elapsed since Schwann made his great discoveries, without the real importance of the new facts having been duly appreciated. This has certainly been essentially due to the great incompleteness of our knowledge with regard to the intimate structure of our tissues which has continued to exist until quite recently, and, as we are sorry to be obliged to confess, still even now prevails with regard to many points of histology to such a degree that we scarcely know in favour of what view to decide.

Especial difficulty has been found in answering the question, from what parts of the body action really proceeds—what parts are active, what passive; and yet it is already quite possible to come to a definitive conclusion upon this point, even in the case of parts the structure of which is still disputed The chief point in this application of histology to pathology is to obtain a recognition of the fact, that the cell is really the ultimate morphological element in which there is any manifestation of life, and that we must not transfer the seat of real action to any point beyond the cell. Before you, I shall have no particular reason to justify myself, if in this respect I make quite a special reservation in favour of life. In the course of these lectures you will be able to convince yourselves that it is almost impossible for any one to entertain more mechanical ideas in particular instances than I am wont to do, when called upon to interpret the individual processes of life. But I think that we must look upon this as certain, that, however much of the more delicate Interchange of matter, which takes place within a cell, may not concern the material structure as a whole, yet the veal action does proceed from the structure as such, and that the living element only maintains its activity as long as it really presents itself to us is an independent whole.

In this question it is of primary importance (and you will excuse my dwelling a little upon this point, as it is one which is still a matter of dispute) that we should determine what is really to be understood by the term cell. Quite at the beginning of the latest phase of histological development, great difficulties sprang up in crowds with regard to this matter. Schwann, as you no doubt recollect, following immediately in the footsteps of Schleiden, interpreted his observations according to botanical standards, so that all the doctrines of vegetable physiology were invoked, in a greater or less degree, to decide questions relating to the physiology of animal bodies. Vegetable cells, however, in the light in which they were at that time universally, and as they are even now also frequently regarded, are structures, whose identity with what we call animal cells cannot be admitted without reserve.

When we speak of ordinary vegetable cellular tissue, we generally understand thereby a tissue, which, in its most simple and regular form is, in a transverse section, seen to be composed of nothing but four- or six-sided, or, if somewhat loose in texture, of roundish or polygonal bodies, in which a tolerably thick, tough wall (*membrane*) is always to be distinguished. If now a single one of these bodies be isolated, a cavity is found, enclosed by this tough, angular, or round wall, in the interior of which very different substances, varying according to circumstances, may be deposited, *e.g.* fat, starch, pigment, albumen (*cell-contents*). But also, quite independently of these local varieties in the contents, we are enabled, by means of chemical investigation, to detect the presence of several different substances in the essential constituents of the cells.

Figure 1. Vegetable cells from the centre of the young shoot of a tuber of *Solarum tuberoum*. *a.* The ordinary appearance of the regularly polygonal, thick-walled cellular tissue. *b.* An isolated cell with finely granular-looking cavity, in which a nucleus with nucleolus is to seen. *c.* The same cell after the addition of water; the contents (protoplasma) have receded from the wall (membrane, capsule). Investing them a peculiar, delicate membrane (primordial utricle) has become visible. *d.* The same cell after a more lengthened exposure to the action of water; the interior cell (protoplasma with the primordial utricle and nucleus) has become quite contracted, and remains attached to the cell-wall (capsule) merely by the means of fine, some of them branching, threads.

The substance which forms the external membrane, and is known under the name of cellulose, is generally found to be destitute of nitrogen, and yields, on the addition of iodine and sulphuric acid, a peculiar, very characteristic, beautiful blue tint. Iodine alone produces no colour; sulphuric add by itself chars. The contents of simple cells, on the other hand, do not turn blue; when the cell is quite a simple one, there appears, on the contrary, after the addition of iodine and sulphuric acid, a brownish or yellowish mass, isolated in the interior of the cell-cavity as a special body (*protoplasma*), around which can be recognised a special, plicated, frequently shriveled membrane (*primordial utricle*) (fig. 1, *c*). Even rough chemical analysis generally detects in the simplest cells, in addition to the non-nitrogenized (external) substance, a nitrogenized internal mass; and vegetable physiology seems, therefore, to have been justified in concluding, that what really constitutes a cell is the presence within a non-nitrogenized membrane of nitrogenized contents differing from it.

It had indeed already long been known, that other things besides existed in the interior of cells, and it was one of the most fruitful of discoveries when Robert Brown detected the *nucleus* in the vegetable cell. But this body was considered to have a more important share in the formation than in the maintenance of cells, because in very many vegetable cells the nucleus becomes extremely indistinct, and in many altogether disappears, whilst the form of the cell is preserved.

These observations were then applied to the consideration of animal tissues, the correspondence of which with those of vegetables Schwann endeavoured to demonstrate. The interpretation, which we have just mentioned as having been put upon the ordinary forms of vegetable cells, served as the starting point. In this, however, as after-experience proved, an error was committed. Vegetable cells cannot, viewed in their entirety, be compared with all animal cells. In animal cells, we find no such distinctions between nitrogenized and non-nitrogenized layers; in all the essential constituents of the cells nitrogenized matters are met with. But there are undoubtedly certain forms in the animal body which immediately recall these forms of vegetable cells, and among them there are none so characteristic as the cells of cartilage, which is, in all its features, extremely different from the other tissues of the animal body, and which, especially on account of its non-vascularity, occupies quite a peculiar position. Cartilage in every respect stands in the closest relation to vegetable tissue. In a well-developed cartilage-cell we can distinguish a relatively thick external layer, within which, upon very close inspection, a delicate membrane, contents, and a nucleus are also to be found. Here, therefore, we have a structure which entirely corresponds with a vegetable cell.

It has, however, been customary with authors, when describing cartilage to call the whole of the structure of which I have just given you a sketch (fig. 2, *a–d*) a cartilage-corpuscle, and in consequence of this having been viewed as analogous to the cells in other parts of animals, difficulties have arisen, by which the knowledge of the true state of the case has been exceedingly obscured. A cartilage-corpuscle, namely, is not, as a whole, a cell, but the external layer, the *capsule*, is the product of a later development (secretion, excretion). In young cartilage it is very thin, whilst the cell also is generally smaller. If we trace the development still farther back, we find in cartilage, also, nothing but simple cells, identical in structure with those which are seen in other animal tissues, and not yet possessing that external secreted layer.

Figure 2. Cartilage-cells, as they occur at the margin of ossification in growing cartilage, quite analogous to vegetable cells (cf. the explanation to fig. 1). *a–c*. In a more advanced stage of development. *d*. Younger form.

You see from this, gentlemen, that the comparison between animal and vegetable cells, which we certainly cannot avoid makings is in general inadmissible, because in most animal tissues no formed elements are found which can be considered as the full equivalents of vegetable cells in the old signification of the word; and because, in particular, the cellulose membrane of vegetable cells does not correspond to the membrane of animal ones, and between this, as containing nitrogen, and the former, as destitute of it, no typical distinction is presented. On the contrary, in both cases we meet with a body essentially of a nitrogenous nature, and, on the whole, similar in composition. The so-called membrane of the vegetable cell is only met with in a few animal tissues, as, for example in cartilage; the ordinary membrane of the animal eel corresponds, as I showed as far back as 1847, to the primordial utricle of the vegetable cell. It is only when we adhere to this view of the matter, when we separate from the cell all that has been added to it by an after-development, that we obtain a simple, homogeneous, extremely monotonous structure, recurring with extraordinary constancy in living organisms. But just this very constancy forms the best criterion of our having before us in this structure one of those really elementary bodies, to be built up of which is eminently characteristic of every living thing—without the pre-existence of which no living forms arise, and to which the continuance and the maintenance of life is intimately attached. Only since our idea of a cell has assumed this severe form—and I am somewhat proud of having always, in spite of the reproach off pedantry, firmly adhered to it—only since that time can it be said that a simple form has been obtained which we can everywhere again expect to find, and which, though different in size and external shape, is yet always identical in its essential constituents.

In such a simple cell we can distinguish dissimilar constituents, and it is important that we should accurately define their nature also.

In the first place, we expect to find a *nucleus* within the cell; and with regard to this nucleus, which has usually a round or oval form, we know that, particularly in the case of young cells, it offers greater resistance to the action of chemical agents than do the external parts off the cell, and that, in spite of the greatest variations in the external form of the cell, it generally maintains its form. The nucleus is accordingly, in cells of all shapes, that part which is the most constantly found unchanged. There are indeed isolated cases, which lie scattered throughout the whole series of facts in comparative anatomy and pathology, in which the nucleus also has a stellate or angular appearance; but these are extremely rare exceptions, and dependent upon peculiar changes which the element has undergone. Generally, it may be said that, as long as the life of the cell has not been brought to a close, as long as cells behave as elements still endowed with vital power, the nucleus maintains a very nearly constant form.

Figure 3. *a.* Hepatic cell. *b.* Spindle-shaped cell from connective tissue. *c.* Capillary vessel. *d.* Somewhat large stellate cell from a lymphatic gland. *e.* Ganglion cell from the cerebellum. The nuclei in every instance similar.

CHAPTER 22
MOLECULES OF LIFE

A cell's major parts are constructed from a few simple molecular building blocks.

INTRODUCTION

C HEMISTS OF THE early 19th century were uncertain if molecules in plants and animals formed by the same rules as those in nonliving systems. This issue was resolved in 1828 by German chemist Friedrich Wöhler (1800–1882), who synthesized urea (a principal component of urine) from common laboratory chemicals. In a contemporaneous letter, Wöhler announced his discovery with humor: "I can no longer, as it were, hold back my chemical urine: and I have to let it out that I can make urea without needing a kidney, whether of man or dog." Wöhler went on to make many other important discoveries, including the first separations of the elements aluminum and beryllium, the discovery of numerous chemical compounds, and the modern method of manufacturing phosphorus.

On the Artificial Production of Urea
by Friedrich Wöhler (1828)

IN A BRIEF earlier communication, printed in Volume III of these Annals, I stated that by the action of cyanogen on aqueous ammonia, besides several other products, there are formed oxalic acid and a crystallizable white substance which is certainly not cyanate of ammonia, but which one nevertheless always obtains when one attempts to combine cyanic acid with ammonia for instance by so-called double decomposition. The fact that in the union of these substances they appear to change their nature, and give rise to a new body, drew my attention anew to this subject, and research gave the unexpected result that by the combination of cyanic acid with ammonia urea is formed, a fact that is the more noteworthy inasmuch as it furnishes an example of the artificial production of an organic, indeed a so-called animal substance, from inorganic materials.

I have already stated that the above-mentioned white crystalline substance is best obtained by the decomposition of cyanate of silver with sal ammoniac solution or of cyanate of lead by aqueous ammonia. In the latter way I prepared for myself the not unimportant amounts employed in this research. I obtained it in colorless, clear crystals often more than an inch long in the form of slender four-sided, dull-pointed prisms.

With caustic potash or chalk this substance evolved no trace of ammonia; with acids it showed none of the breakdown phenomena of cyanic acid salts, namely, evolution of carbonic acid and cyanic acid; neither would it precipitate lead and silver salts as genuine cyanic acid salts do; it could, therefore, contain neither cyanic acid nor ammonia as such. Since I found that by the last-named method of preparation no other product was formed and that the lead oxide was separated in a pure form, I imagined that an organic substance might arise by the union of cyanic acid with ammonia, possibly a substance like a vegetable salifiable base [an alkaloid]. I therefore made some experiments from this point of view on the behavior of the crystalline substance with acids. It was, however, indifferent to them, nitric acid excepted this, when added to a concentrated solution of the substance, produced at once a precipitate of glistening scales. After these had been purified by several recrystallizations, they showed a very acid character, and I was already inclined to take the compound for a peculiar acid, when i found that after neutralization with bases it gave salts of nitric acid, from which the crystallizable substance could be extracted again with alcohol, with all the characteristics it had before the addition of nitric acid. This similarity to urea in behavior induced me to carry out comparative experiments with completely pure urea isolated from urine, from which it was plainly apparent that urea and this crystalline substance, or cyanate of ammonia, if one can so call it, are completely identical compounds.

I will describe the properties of this artificial urea no further, since it coincides perfectly with that of urea from urine, according to the accounts of Proust, Prout and others, to be found in their writings, and I will mention only the fact, not specified by them, that both natural and artificial urea, on distillation evolve first large amounts of carbonate of ammonia, and then give off to a remarkable extent the sharp, acetic acic cyanic acid, exactly as I found in the distillation of cyanate of uric acid, and especially of the mercury salt of uric acid. In the urea, another white, apparently distinct substance also appears, with the examination of which I am still occupied.

But if the combination of cyanic acid and ammonia actually gives merely urea, it must have exactly the composition allotted to cyanate of ammonia by calculation from my formula for the cyanates; and this is, in fact, the case if one atom of water is added to cyanate of ammonia, as all ammonium salts contain water, and if Prout's analysis of urea is taken as the most correct.

According to him urea consists of:

		Atoms
Nitrogen	46.650	4
Carbon	19.975	2
Hydrogen	6.670	8
Oxygen	26.650	2
	99.875 [sic]	

But cyanate of ammonia would consist of 56.92 cyanic acid, 28.14 ammonia, and 14.74 water, which for the separate elements gives:

		Atoms
Nitrogen	46.78	4
Carbon	20.19	2
Hydrogen	6.59	8
Oxygen	26.24	2
	99.80	

One would have been able to reckon beforehand that cyanate of ammonia with 1 atom of water has the same composition as: urea, without having discovered by experiment the formation of urea from cyanic acid and ammonia. By the combustion of cyanic acid with copper oxide one obtains 2 volumes of carbonic acid and 1 volume of nitrogen, but by the combustion of cyanate of ammonia one must obtain equal volumes of these gases, which proportion also holds for urea, as Prout actually found.

CHAPTER 23
CLASSICAL AND MODERN GENETICS

All living things use the same genetic code to guide chemical reactions in every cell.

INTRODUCTION

G ENETICS, THE STUDY of how organisms pass biological information from one generation to the next, is transforming our world at every level from individuals to society as a whole. In this chapter and the next we explore key discoveries in the science and technology of genetics. In the following three excerpts we present pivotal findings in the three great stages of genetic science: classic genetics (the study of organisms), cellular genetics (the study of chromosomes), and molecular genetics (the study of DNA).

Gregor Mendel (1822–1884) is honored as the founder of the science of genetics, yet his lasting contributions were completed during an 8-year period in a life devoted principally to the administration of church business in the city of Brno in what was then part of the Austro-Hungarian empire. Mendel applied meticulous horticultural techniques and record keeping to investigate the cross-breeding of different varieties of pea plants. These experiments led him to three laws of heredity: there exist "atoms of inheritance" (what we now call genes) that pass from parents to offspring; each parent contributes half of these atoms of inheritance; and some traits are dominant while others are recessive.

Experiments in Plant Hybridization[1]
by Gregor Mendel (1866)

[1] INTRODUCTORY REMARKS

EXPERIENCE OF ARTIFICIAL fertilization, such as is effected with ornamental plants in order to obtain new variations in color, has led to the experiments which will here be discussed. The striking regularity with which the same hybrid forms always reappeared whenever fertilization took place between the same species induced further experiments to be undertaken, the object of which was to follow up the developments of the hybrids in their progeny.

To this object numerous careful observers, such as Kölreuter, Gärtner, Herbert, Lecoq, Wichura and others, have devoted a part of their lives with inexhaustible perseverance. Gärtner especially in his work *Die Bastarderzeugung im Pflanzenreiche*, has recorded very valuable observations; and quite recently Wichura published the results of some profound investigations into the hybrids of the Willow. That, so far, no generally applicable law governing the formation and development of hybrids has been successfully formulated can hardly be wondered at by anyone who is acquainted with the extent of the task, and can appreciate the difficulties with which experiments of this class have to contend. A final decision can only be arrived at when we shall have before us the results of *detailed experiments* made on plants belonging to the most diverse orders.

Those who survey the work done in this department will arrive at the conviction that among all the numerous experiments made, not one has been carried out to such an extent and in such a way as to make it possible to determine the number of different forms under which the offspring of the hybrids appear, or to arrange these forms with certainty according to their separate generations, or definitely to ascertain their statistical relations.

It requires indeed some courage to undertake a labor of such far-reaching extent; this appears, however, to be the only right way by which we can finally reach the solution of a question the importance of which cannot be overestimated in connection with the history of the evolution of organic forms.

The paper now presented records the results of such a detailed experiment. This experiment was practically confined to a small plant group, and is now, after eight years' pursuit, concluded in all essentials. Whether the plan upon which the separate experiments were

---

1 Translated from the German: "Versuche über Plflanzenhybriden."

conducted and carried out was the best suited to attain the desired end is left to the friendly decision of the reader.

[2] SELECTION OF THE EXPERIMENTAL PLANTS

The value and utility of any experiment are determined by the fitness of the material to the purpose for which it is used, and thus in the case before us it cannot be immaterial what plants are subjected to experiment and in what manner such experiment is conducted.

The selection of the plant group which shall serve for experiments of this kind must be made with all possible care if it be desired to avoid from the outset every risk of questionable results.

The experimental plants must necessarily:

1.   Possess constant differentiating characteristics
2.   The hybrids of such plants must, during the flowering period, be protected from the influence of all foreign pollen, or be easily capable of such protection.
3.   The hybrids and their offspring should suffer no marked disturbance in their fertility in the successive generations.

Accidental impregnation by foreign pollen, if it occurred during the experiments and were not recognized, would lead to entirely erroneous conclusions. Reduced fertility or entire sterility of certain forms, such as occurs in the offspring of many hybrids, would render the experiments very difficult or entirely frustrate them. In order to discover the relations in which the hybrid forms stand towards each other and also towards their progenitors it appears to be necessary that all member of the series developed in each successive generations should be, *without exception*, subjected to observation.

At the very outset special attention was devoted to the *Leguminosae* on account of their peculiar floral structure. Experiments which were made with several members of this family led to the result that the genus *Pisum* was found to possess the necessary qualifications. ...

In all, 34 more or less distinct varieties of Peas were obtained from several seedsmen and subjected to a two year's trial. In the case of one variety there were noticed, among a larger number of plants all alike, a few forms which were markedly different. These, however, did not vary in the following year, and agreed entirely with another variety obtained from the same seedsman; the seeds were therefore doubtless merely accidentally mixed. All the other varieties yielded perfectly constant and similar offspring; at any rate, no essential difference was observed during two trial years. For fertilization 22 of these were selected and cultivated during the whole period of the experiments. They remained constant without any exception. ...

[3] DIVISION AND ARRANGEMENT OF THE EXPERIMENTS

If two plants which differ constantly in one or several characters be crossed, numerous experiments have demonstrated that the common characters are transmitted unchanged to the hybrids and their progeny; but each pair of differentiating characters, on the other hand, unite in the hybrid to form a new character, which in the progeny of the hybrid is usually variable. The object of the experiment was to observe these variations in the case of each pair of differentiating characters, and to deduce the law according to which they appear in successive generations. The experiment resolves itself therefore into just as many separate experiments are there are constantly differentiating characters presented in the experimental plants.

The various forms of Peas selected for crossing showed differences in length and color of the stem; in the size and form of the leaves; in the position, color, size of the flowers; in the length of the flower stalk; in the color, form, and size of the pods; in the form and size of the seeds; and in the color of the seed-coats and of the albumen (endosperm). Some of the characters noted do not permit of a sharp and certain separation, since the difference is of a "more or less" nature, which is often difficult to define. Such characters could not be utilized for the separate experiments; these could only be applied to characters which stand out clearly and definitely in the plants. Lastly, the result must show whether they, in their entirety, observe a regular behavior in their hybrid unions, and whether from these facts any conclusion can be reached regarding those characters which possess a subordinate significance in the type.

The characters which were selected for experiment relate:

1.   To the *difference in the form of the ripe seeds*. These are either round or roundish, the depressions, if any, occur on the surface, being always only shallow; or they are irregularly angular and deeply wrinkled (P. quadratum).

2.   To the *difference in the color of the seed albumen* (endosperm). The albumen of the ripe seeds is either pale yellow, bright yellow and orange colored, or it possesses a more or less intense green tint. This difference of color is easily seen in the seeds as their coats are transparent.

3.   To the *difference in the color of the seed-coat*. This is either white, with which character white flowers are constantly correlated; or it is gray, gray-brown, leather-brown, with or without violet spotting, in which case the color of the standards is violet, that of the wings purple, and the stem in the axils of the leaves is of a reddish tint. The gray seed-coats become dark brown in boiling water.

4.   To the *difference in the form of the ripe pods*. These are either simply inflated, not contracted in places; or they are deeply constricted between the seeds and more or less wrinkled (P. saccharatum).

5.   To the *difference in the color of the unripe pods*. They are either light to dark green, or vividly yellow, in which coloring the stalks, leaf-veins, and calyx participate.

6.   To the *difference in the position of the flowers*. They are either axial, that is, distributed along the main stem; or they are terminal, that is, bunched at the top of the stem and arranged almost in a false umbel; in this case the upper part of the stem is more or less widened in section (P. umbellatum).

7.   To the *difference in the length of the stem.* The length of the stem is very various in some forms; it is, however, a constant character for each, in so far that healthy plants, grown in the same soil, are only subject to unimportant variations in this character. In experiments with this character, in order to be able to discriminate with certainty, the long axis of 6 to 7 ft. was always crossed with the short one of 3/4 ft. to 1 [and] 1/2 ft. ...

From a larger number of plants of the same variety only the most vigorous were chosen for fertilization. Weakly plants always afford uncertain results, because even in the first generation of hybrids, and still more so in the subsequent ones, many of the offspring either entirely fail to flower or only form a few and inferior seeds. ...

The plants were grown in garden beds, a few also in pots, and were maintained in their natural upright position by means of sticks, branches of trees, and strings stretched between. For each experiment a number of pot plants were placed during the blooming period in a greenhouse, to serve as control plants for the main experiment in the open as regards possible disturbance by insects. ...

[4] THE FORMS OF THE HYBRIDS

Experiments which in previous years were made with ornamental plants have already affording evidence that the hybrids, as a rule, are not exactly intermediate between the parental species. With some of the more striking characters, those, for instance, which relate to the form and size of the leaves, the pubescence of the several parts, etc., the intermediate, indeed, is nearly always to be seen; in other cases, however, one of the two parental characters is so preponderant that it is difficult, or quite impossible, to detect the other in the hybrid.

This is precisely the case with the Pea hybrids. In the case of each of the 7 crosses the hybrid-character resembles that of one of the parental forms so closely that the other either escapes observation completely or cannot be detected with certainty. This circumstance is of great importance in the determination and classification of the forms under which the offspring of the hybrids appear. Henceforth in this paper those characters which are transmitted entire, or almost unchanged in the hybridization, and therefore in themselves constitute the characters of the hybrid, are termed the *dominant*, and those which become latent in the process *recessive.* The expression "recessive" has been chosen because the characters thereby designated withdraw or entirely disappear in the hybrids, but nevertheless reappear unchanged in their progeny, as will be demonstrated later on.

It was furthermore shown by the whole of the experiments that it is perfectly immaterial whether the dominant character belongs to the seed plant or to the pollen plant; the form of the hybrid remains identical in both cases. ...

Of the differentiating characters which were used in the experiments the following are dominant:
1.   The round or roundish form of the seed with or without shallow depressions.

2. The yellow coloring of the seed albumen.

3. The gray, gray-brown, or leather brown color of the seed-coat, in association with violet-red blossoms and reddish spots in the leaf axils.

4. The simply inflated form of the pod.

5. The green coloring of the unripe pod in association with the same color of the stems, the leaf-veins and the calyx.

6. The distribution of the flowers along the stem.

7. The greater length of stem. ...

[5] THE FIRST GENERATION FROM THE HYBRIDS

In this generation there reappear, *together with the dominant* characters, also the *recessive* ones with their peculiarities fully developed, and this occurs in the definitely expressed average proportion of 3:1, so that among each 4 plants of this generation 3 display the dominant character and one the recessive. This relates without exception to all the characters which were investigated in the experiments. The angular wrinkled form of the seed, the green color of the albumen, the white color of the seed-coats and the flowers, the constriction of the pods, the yellow color of the unripe pod, of the stalk, of the calyx, and of the leaf venation, the umbel-like form of the inflorescence, and the dwarfed stem, all reappear in the numerical proportion given, without any essential alteration. *Transitional forms were not observed in any experiment.* ...

• Expt. 3: Color of the seed-coats. Among 929 plants, 705 bore violet-red flowers and gray-brown seed-coats; 224 had white flowers and white seed-coats, giving the proportion 3.15:1.

• Expt. 4: Form of pods. Of 1181 plants, 882 had them simply inflated, and in 299 they were constricted. Resulting ratio, 2.95:1. ...

• Expt. 7: Length of stem. Out of 1064 plants, in 787 cases the stem was long, and in 277 short. Hence a mutual ratio of 2.84:1. ...

If now the results of the whole of the experiments be brought together, there is found, as between the number of forms with the dominant and recessive characters, an average ratio of 2.98:1, or 3:1. ...

[6] THE SECOND GENERATION FROM THE HYBRIDS

Those forms which in the first generation exhibit the recessive character do not further vary in the second generation as regards this character; they remain *constant* in their offspring.

It is otherwise with those which possess the dominant character in the first generation. Of these *two*-thirds yield offspring which display the dominant and recessive characters in the proportion of 3:1, and thereby show exactly the same ratio as the hybrid forms, while only *one*-third remains with the dominant character constant.

The separate experiments yielded the following results:

• Expt. 1: Among 565 plants which were raised from round seeds of the first generation, 193 yielded round seeds only, and remained therefore constant in this character; 372, however, gave both round and wrinkled seeds, in the proportion of 3:1. The number of the hybrids, therefore, as compared with the constants is 1.93:1.

• Expt. 2: Of 519 plants which were raised from seeds whose albumen was of yellow color in the first generation, 166 yielded exclusively yellow, while 353 yielded yellow and green seeds in the proportion of 3:1. There resulted, therefore, a division into hybrid and constant forms in the proportion of 2.13:1. …

In each of these experiments a certain number of the plants came constant with the dominant character. For the determination of the proportion in which the separation of the forms with the constantly persistent character results, the two first experiments are especially important, since in these a larger number of plants can be compared. The ratios 1.93:1 and 2.13:1 gave together almost exactly the average ratio of 2:1. … *The average ratio of 2:1 appears, therefore, as fixed with certainty.* It is therefore demonstrated that, of those forms which posses the dominant character in the first generation, two-thirds have the hybrid-character, while one-third remains constant with the dominant character.

The ratio 3:1, in accordance with which the distribution of the dominant and recessive characters results in the first generation, resolves itself therefore *in all experiments into the ratio of 2:1:1*, if the dominant character be differentiated according to its significance as a hybrid-character or as a parental one. Since the members of the first generation spring directly from the seed of the hybrids, *it is now clear that the hybrids form seeds having one or other of the two differentiating characters, and of these one-half develop again the hybrid form, while the other half yield plants which remain constant and receive the dominant or the recessive characters in equal numbers.*

The Mechanism of Mendelian Heredity
by T. H. Morgan, A. H. Sturtevant, H. J. Muller, and C. B. Bridges (1915)

THOMAS HUNT MORGAN (1866–1945) and his students at Columbia University, Alfred H. Sturtevant (1891–1970), Hermann J. Muller (1890–1967), and Calvin B. Bridges (1889–1938)—collectively known as the "fly boys," studied the traits of the fast-breeding fruit fly, *Drosophila ampelophila*. They discovered that certain traits of the flies were linked, and concluded that such pairing revealed traits that were associated with one of the fly's four chromosomes. They concluded that the more frequently two traits are linked, the closer the governing structures must lie on a chromosome. In this way, they were able to map out the relative position of genes on the chromosomes.

* * *

PREFACE

From ancient times heredity has been looked upon as one of the central problems of biological philosophy. It is true that this interest was largely speculative rather than empirical. But since Mendel's discovery of the fundamental law of heredity in 1865, or rather since its re-discovery in 1900, a curious situation has begun to develop. The students of heredity calling themselves geneticists have begun to draw away from the traditional fields of zoology and botany, and have concentrated their attention on the study of Mendel's principles and their later developments. The results of these investigators appear largely in special journals. Their terminology is often regarded by other zoologists as something barbarous,—outside the ordinary routine of their profession. The tendency is to regard genetics as a subject for specialists instead of an all-important theme of zoology and botany. No doubt this is but a passing phase; for biologists can little afford to hand over to a special group of investigators a part of their field that is and always will be of vital import. It would be as unfortunate for all biologists to remain ignorant of the modern advances in the study of heredity as it would be for the geneticists to remain unconcerned as to the value for their own work of many special fields of biological inquiry. What is fundamental in zoology and botany is not so extensive, or so intrinsically difficult, that a man equipped for his profession should not be able to compass it.

In the following pages we have attempted to separate those questions that seem to us significant from that which is special or merely technical. We have, of course, put our own interpretation on the facts, and while this may not be agreed to on all sides, yet we believe that in what is essential we have not departed from the point of view that is held by many of our co-workers at the present time. Exception may perhaps be taken to the emphasis we have laid on the chromosomes as the material basis of inheritance. Whether we are right here, the future—probably a very near future—will decide. But it should not pass unnoticed that even if the chromosome theory be denied, there is no result dealt with in the following pages that may not be treated independently of the chromosomes; for, we have made no assumption concerning heredity that cannot also be made abstractly without the chromosomes as bearers of the postulated hereditary factors. Why then, we are often asked, do you drag in the chromosomes? Our answer is that since the chromosomes furnish exactly the kind of mechanism that the Mendelian laws call for; and since there is an ever-increasing body of information that points clearly to the chromosomes as the bearers of the Mendelian factors, it would be folly to close one's eyes to so patent a relation. Moreover, as biologists, we are interested in heredity not primarily as a mathematical formulation but rather as a problem concerning the cell, the egg, and the sperm.

CHAPTER I. MENDELIAN SEGREGATION AND THE CHROMOSOMES

Mendel's law was announced in 1865. Its fundamental principle is very simple. *The units contributed by two parents separate in the germ cells of the offspring without having had any influence on each other.* For example, in a cross between yellow-seeded and green-seeded peas, one parent contributes to the offspring a unit for yellow and the other parent contributes a unit for green. These units separate in the ripening of the germ cells of the offspring so that half of the germ cells are yellow bearing and half are green bearing. This separation occurs both in the eggs and in the sperm.

Mendel did not know of any mechanism by which such a process could take place. In fact, in 1865 very little was known about the ripening of the germ cells. But in 1900, when Mendel's long-forgotten discovery was brought to light once more, a mechanism had been discovered that fulfils exactly the Mendelian requirements of pairing and separation.

The sperm of every species of animal or plant carries a definite number of bodies called chromosomes. The egg carries the same number. Consequently, when the sperm unites with the egg, the fertilized egg will contain the double number of chromosomes. For each chromosome contributed by the sperm there is a corresponding chromosome contributed by the egg, i.e., there are two chromosomes of each kind, which together constitute a pair.

When the egg divides (Fig. 1, *a–d*) every chromosome splits into two chromosomes, and these daughter chromosomes then move apart, going to opposite poles of the dividing cell (Fig. 1, *c*). Thus each daughter cell (Fig. 1, *d*) receives one of the daughter chromosomes formed from each original chromosome. The same process occurs in all cell divisions, so that all the cells of the animal or plant come to contain the double set of chromosomes.

The germ cells also have at first the double set of chromosomes, but when they are ready to go through the last stages of their transformation into sperm or eggs the chromosomes unite in pairs (Fig. 1, *e*). Then follows a different kind of division (Fig. 1, *f*), at which the chromosomes do not split, but the members of each pair of chromosomes separate and each member goes into one of the daughter cells (Fig. 1, *g, h*). As a result each mature germ cell receives one or the other member of every pair of chromosomes and the number is reduced to half. Thus the behaviour of the chromosomes parallels the behaviour of the Mendelian units, for in the germ cells each unit derived from the father separates from the corresponding unit derived from the mother. These units will henceforth be spoken of as factors; the two factors of a pair are called allelomorphs of each other. Their separation in the germ cells is called segregation.

The possibility of explaining Mendelian phenomena by means of the manoeuvres of the chromosomes seems to have occurred to more than one person, but Sutton was the first to present the idea in the form in which we recognize it today. Moreover, he not only called attention to the fact above mentioned, that both chromosomes and hereditary factors undergo segregation, but showed that the parallelism between their methods of distribution goes even further than this. Mendel had found that when the inheritance of more than one pair of factors is followed, the different pairs of factors segregate independently of one another. Thus in a cross of a pea having both green seeds and tall stature with a pea having yellow seeds and short stature, the fact that a germ cell receives a particular member of one pair (e.g. yellow) does not determine which member of the other pair it receives; it is as likely to receive the tall as the short. Sutton pointed out that in the same way the segregation of one pair of chromosomes is probably independent of the segregation of the other pairs.

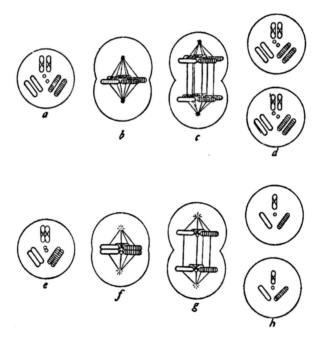

Figure 1. In the upper line, four stages in the division of the egg (or of a body-cell) are represented. Every chromosome divides when the cell divides. In the lower line the "reduction division" of a germ cell, after the chromosomes have united in pairs, is represented. The members of each of the four pairs of chromosomes separate from each other at this division.

It was obvious from the beginning, however, that there was one essential requirement of the chromosome view, namely, that all the factors carried by the *same* chromosome should tend to remain together. Therefore, since the number of inheritable characters may be large in comparison with the number of pairs of chromosomes, we should expect actually to find not only the independent behaviour of pairs, but also cases in which characters are linked together in groups in their inheritance. Even in species where a limited number of Mendelian units are known, we should still expect to find some of them in groups.

In 1906 Bateson and Punnett made the discovery of linkage, which they called gametic coupling. They found that when a sweet pea with factors for purple flowers and long pollen grains was crossed to a pea with factors for red flowers and round pollen grains, the two factors that came from the same parent tended to be inherited together. Here was the first case that gave the sort of result that was to be expected if factors were in chromosomes, although this relation was not pointed out at the time. In the same year, however, Lock called attention to the possible relation between the chromosome hypothesis and linkage.

In other groups a few cases of coupling became known, but nowhere had the evidence been sufficiently ample or sufficiently studied to show how frequently coupling occurs. Since 1910, however, in the fruit fly *Drosophila ampelophila*, a large number of new characters have appeared by mutation, and so rapidly does the animal reproduce that in a relatively short time the inheritance of more than a hundred characters has been studied. It became evident very soon that these characters are inherited in groups. There is one great group of characters that are sex linked. There are two other groups of characters slightly greater in number. Finally a character appeared that did not belong to any of the other groups, and a year later still another character appeared that was linked to the last one but was independent of all the other groups. Hence there are four groups of characters in *Drosophila*. …

Figure 2. Diagram of female group (duplex) of chromosomes of *Drosophila ampelophila* showing four pairs of chromosomes. The members of each pair are usually found together, as here.

The four pairs of chromosomes of … *Drosophila* are shown in Fig. 2. There are three pairs of large chromosomes and one pair of small chromosomes. …

In *Drosophila*, then, there is a numerical correspondence between the number of hereditary groups and the number of the chromosomes. Moreover, the size relations of the groups and of the chromosomes correspond. ...

In most animals and plants the number of chromosomes is higher than in *Drosophila*, and the number of pairs of factors that may show independent assortment is, in consequence, increased. In the snail, *Helix hortensis*, the half number of the chromosomes is given as 22; in the potato beetle 18; in man, probably 24; in the mouse 20; in cotton 28;...in the garden pea 7; in the nightshade 36; in tobacco 24; in the tomato 12; in wheat 8. If 20 pairs of chromosomes are present there will be over one million possible kinds of germ cells in the F_1 hybrid. The number of combinations that two such sets of germ cells may produce through fertilization is enormously greater. From this point of view we can understand the absence of identical individuals in such mixed types as the human race. The chance of identity is still further decreased since in addition there may be very large numbers of factors within each chromosome.

A Structure for Deoxyribose Nucleic Acid
by James D. Watson and Francis H. C. Crick (1953)

THE DISCOVERY THAT chromosomes are made largely of deoxyribose nucleic acid, or DNA, led most biologists to conclude that DNA carried the genetic information. Convincing proof came with the determination of the DNA "double helix" structure by American James D. Watson (1928–) and Francis H. C. Crick (1916–2004). Relying heavily on x-ray data collected by British crystallographer Rosalind Franklin (1920–1958) and the insights of New Zealand-born biophysicists Maurice Wilkins (1916–2004), they solved the elegant and complex structure just ahead of several competing research groups. Their concise 1953 *Nature* paper is one of the most influential scientific discoveries of the 20th century.

* * *

We wish to suggest a structure for the salt of deoxyribose nucleic acid (D.N.A.). This structure has novel features which are of considerable biological interest.

A structure for nucleic acid has already been proposed by Pauling and Corey.[1] They kindly made their manuscript available to us in advance of publication. Their model consists of three intertwined chains, with the phosphates near the fibro axis, and the bases on the outside. In our opinion, this structure is unsatisfactory for two reasons: (1) We believe that, the material which gives the X-ray diagrams is the salt, not the free acid. Without the acidic hydrogen atoms it is not clear what forces would hold the structure together, especially as the negatively charged phosphates near the axis will repel each other. (2) Some of the van der Waals distances appear to be too small.

Another three-chain structure has also been suggested by Fraser (in the press). In his model the phosphates are on the outside and the bases on the inside, linked together by hydrogen bonds. This structure as described is rather ill-defined, and for this reason we shall not comment on it.

We wish to put forward a radically different structure for the salt of deoxyribose nucleic acid. This structure has two helical chains each coiled round the same axis (see diagram). ...

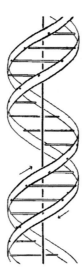

This figure is purely diagrammatic. The two ribbons symbolize the two phosphate-sugar chains, and the horizontal rods the pairs of bases holding the chains together. The vertical line marks the fibre axis.

The novel feature of the structure is the manner in which the two chains are held together by the purine and pyrimidine bases. The planes of the bases are perpendicular to the fibro axis. They are joined together in pairs, a single base from one chain being hydrogen-bonded to a single base from the other chain, so that the two lie side by side with identical z-coordinates. One of the pair must be a purine and the other a pyrimidine for bonding to occur. The hydrogen bonds are made as follows: purine position 1 to pyrimidine position 1; purine position 6 to pyrimidine position 6.

If it is assumed that the bases only occur in tho structure in the most, plausible tautomeric forms (that is, with the keto rather than the enol configurations) it is found that only specific pairs of bases can bond together. These pairs are: adenine (purine) with thymine (pyrimidine), and guanine (purine) with cytosine (pyrimidine).

In other words, if an adenine forms one member of a pair, on either chain, then on these assumptions the other member must be thymine; similarly for guanine and cytosine. The sequence of bases on a single chain does not appear to be restricted in anyway. However, if only specific pairs of bases can be formed, it follows that if the sequence of bases on one chain is given, then the sequence on the other chain is automatically determined.

It has been found experimentally[3-4] that the ratio of the amounts of adenine to thymine, and the ratio of guanine to cytosine, are always very close to unity for deoxyribose nucleic acid.

It is probably impossible to build this structure with a ribose sugar in place of the deoxyribose, as the extra oxygen atom would make too close a van der Waals contact.

The previously published X-ray data[5-6] on deoxyribose nucleic acid are insufficient for a rigorous test of our structure. So far as we can tell, it is roughly compatible with the experimental data, but it must be regarded as unproved until it has been checked against more exact results. Some of these are given in the following communications. We were not aware of the details of the results presented there when we devised our structure, which rests mainly though not entirely on published experimental data and stereochemical arguments.

It has not escaped our notice that the specific pairing we have postulated immediately-suggests a possible copying mechanism for the genetic material.

Full details of the structure, including the conditions assumed in building it, together with a set of co-ordinates for the atoms, will be published elsewhere.

We are much indebted to Dr. Jerry Donohue for constant advice and criticism, especially on interatomic distances. We have also been stimulated by a knowledge of the general nature of the unpublished experimental results and ideas of Dr. M. H. F. Wilkins, Dr. R. E. Franklin and their co-workers at King's College, London. One of us (J.D.W.) has been aided by a fellowship from the National Foundation for Infantile Paralysis.

J. D. Watson

F. H. C. Crick

Medical Research Council Unit for the Study of the Molecular Structure of Biological Systems, Cavendish Laboratory, Cambridge. April 2.

1. Pauling, L. and Corey. R. B.. *Nature*, 171, 346 (1953); *Proc. U.S. Nat. Acad. Sci.*, 39. 84 (1953).

2. Furberg, S., *Acta Chem. Scand.*, 6. 634 (1952).

3. Chargaff, E., for references see Zamenhof, S., Brawerman, G. and Chargaff. E. *Biochim. et Biophys. Acta*. 9. 402 (1952).

4. Wyatt. G. R., *J. Gen. Physiol.* 36. 201 (1952).

5. Astbury, W. T., Symp. Soc. Exp. Biol. 1, Nucleic Acid, 66 (Camb.Univ. Press. 1947).

6. Wilkins, M. H. F., and Randall, J. T., *Biochim. et Biophys. Acta*. 10, 192 (1953).

CHAPTER 24
THE NEW SCIENCE OF LIFE

Our new understanding of genetic mechanisms is leading to enormous advances in medicine and other aspects of our lives.

INTRODUCTION

THE HALLMARK OF modern molecular genetics is the ability to manipulate the genetic molecule, DNA. The following two articles describe transformational technological genetic advances—the polymerase chain reaction and mammal cloning. These two papers contrast with most of the others in this volume for at least three reasons. First, unlike most of the other works, each of these contributions has many authors—a sign of the complexity of the work and specialization of modern science. Second, each of these papers represents technologies that have tremendous medical implications, and thus are worth a lot of money. Thus, in each case patent applications preceded the scientific paper. And, finally, both of these biotech articles are exceptionally technical—so much so that few of the Ph.D. scientist readers of *Science* and *Nature* would have been able to follow the details. We've edited out some of that detail, but if these articles seem difficult to wade through, don't be too discouraged.

The ability to make millions of copies of a single piece of DNA—the polymerase chain reaction, or PCR—has revolutionized many aspects of science and technology, including genetic engineering, medical diagnosis, paternity tests, and crime scene investigation. PCR was invented by California molecular biologist (and avid surfer) Kary B. Mullis (1944–), who shared the 1993 Chemistry Nobel for his discovery. PCR works by putting a target piece of DNA and lots of DNA building blocks into a solution, which is first heated (to separate the two long halves of the DNA double helix) and then cooled (to induce each half-DNA strand to form a complete double helix). Thus, the number of DNA strands doubles with every temperature cycle: 20 cycles can yield a million copies. After filing patent applications in 1985, Mullis and colleagues submitted a paper to *Nature* describing the revolutionary process, but that paper was rejected! Eager to make sure they got scientific credit, they quickly tacked the PCR method into another paper on a new way to diagnose the genetic disease sickle-cell anemia from fetal cells using the enzyme β-globin. That paper, excerpted here, was quickly published by *Science* in the last issue of 1985.

Enzymatic Amplification of β-Globin Genomic Sequences and Restriction Site Analysis for Diagnosis of Sickle Cell Anemia

by Randall K. Saiki, Stephen Scharf, Fred Faloona, Kary B. Mullis, Glenn T. Horn, Henry A. Erlich, and Norman Arnheim (1985)

ABSTRACT. TWO NEW methods were used to establish a rapid and highly sensitive prenatal diagnostic test for sickle cell anemia. The first involves the primer-mediated enzymatic amplification of specific β-globin target sequences in genomic DNA, resulting in the exponential increase (220,000 times) of target DNA copies. In the second technique, the presence of the β^A and β^S alleles is determined by restriction endonuclease digestion of an end-labeled oligonucleotide probe hybridized in solution to the amplified β-globin sequences. The β-globin genotype can be determined in less than 1 day on samples containing significantly less than 1 microgram of genomic DNA.

Recent advances in recombinant DNA technology have made possible the molecular analysis and prenatal diagnosis of several human genetic diseases. Fetal DNA obtained by aminocentesis or chorionic villus sampling can be analyzed by restriction enzyme digestion, with subsequent electrophoresis. Southern transfer, and specific hybridization to cloned gene or oligonucleotide probes. With polymorphic DNA markers linked genetically to a specific disease locus, segregation analysis must be carried out with restriction fragment length polymorphisms (RFLP's) found to be informative by examining DNA from family members.

Many of the hemoglobinopathies, however, can be detected by more direct methods in which analysis of the fetus alone is sufficient for diagnosis. For example, the diagnosis of hydrops fetalis (homozygous α-thalassemia) can be made by documenting the absence of any α-globin genes by hybridization with an α-globin probe. Homozygosity for certain β-thalassemia alleles can be determined in Southern transfer experiments by using oligonucleotide probes that form stable duplexes with the normal β-globin gene sequence but form unstable hybrids with specific mutants.

Sickle cell anemia can also be diagnosed by direct analysis of fetal DNA. This disease results from homozygosity of the sickle-cell allele (β^S) at the β-globin gene locus. The S allele differs from the wild-type allele (β^A) by substitution of an A in the wild-type to a T at the second position of the sixth codon of the 3 chain gene, resulting in the replacement of a glutamic acid by a valine in the expressed protein. For the prenatal diagnosis of sickle cell anemia, DNA obtained by amniocentesis or chorionic villus sampling can be treated with a restriction endonuclease (for example, Dde I and Mst II) that recognizes a sequence altered by the β^S mutation. This generates β^A- and β^S-specific restriction fragments that can be resolved by Southern transfer and hybridization with a β-globin probe.

We have developed a procedure for the detection of the sickle cell mutation that is very rapid and is at least two orders of magnitude more sensitive than standard Southern blotting. There, are two special features to this protocol. The first is a method for amplifying specific β-globin DNA sequences with the use of oligonucleotide primers and DNA polymerase. The second is the analysis of the β-globin genotype by solution hybridization of the amplified DNA with a specific oligonucleotide probe and subsequent digestion with a restriction endonuclease. These two techniques increase the speed and sensitivity, and lessen the complexity of prenatal diagnosis for sickle cell anemia; they may also be generally applicable to the diagnosis of other genetic diseases and in the use of DNA probes for infectious disease diagnosis.

SEQUENCE AMPLIFICATION BY POLYMERASE CHAIN REACTION

We use a two-step procedure for determining the p-globin genotype of human genomic DNA samples. First, a small portion of the β-globin gene sequence spanning the polymorphic Dde I restriction site diagnostic of the β^A allele is amplified. Next, the presence or absence of the Dde I restriction site in the amplified DNA sample is determined by solution hybridization with an end-labeled complementary oligomer followed by restriction endonuclease digestion, electrophoresis, and autoradiography.

The β-globin gene segment was amplified by the polymerase chain reaction (PCR) procedure of Mullis and Faloona in which we used two 20-base oligonucleotide primers that flank the region to be amplified. One primer, PC04, is complementary to the (+)-strand and the other, PC03, is complementary to the (-)-strand. The annealing of PC04 to the (+)-strand of denatured genomic DNA followed by extension with the Klenow fragment of *Escherichia coli* DNA polymerase I and deoxynucleotide triphosphates results in the synthesis of a (-)-strand fragment containing the target sequence. At the same time, a similar reaction occurs with PC03, creating a new (+)-strand. Since these newly synthesized DNA strands are themselves template for the PCR primers, repeated cycles of denaturation, primer annealing, and extension result in the exponential accumulation of the 110-base pair region defined by the primers.

… Samples of DNA (1 μg) were amplified for 20 cycles and a fraction of each sample, equivalent to 36 ng of the original DNA, was subjected to alkaline gel electrophoresis and transferred to a nylon filter. The filter was then hybridized with a ^{32}P-labeled 40-base oligonucleotide probe, RS06, which is complementary to the target sequence but not to the PCR primers. The results, after a 2-hour autoradiographic exposure, show that a fragment hybridizing with the RS06 probe. …

The application of the PCR method to prenatal diagnosis does not necessarily depend on a polymorphic restriction site or on the use of radioactive probes. In fact, the significant amplification of target sequences achieved by the PCR method allows the use of nonisotopically labeled probes. Amplified target sequences could also be analyzed by a number of other procedures including those involving the hybridization of small labeled oligomers which will form stable duplexes only if perfectly matched and the recently reported method based on the electrophoretic shifts of duplexes with base pair mismatches. The ability of the PCR procedure to amplify a target DNA segment in genomic DNA raises the possibility that its use may extend beyond that of prenatal diagnosis to other areas of molecular biology.

Viable Offspring Derived from Fetal and Adult Mammalian Cells

by I. Wilmut, A. E. Schnieke, J. McWhlr, A. J. Kind, and K. H. S. Campbell (1997)

CLONING IS ONE of the most scientifically (not to mention ethically) challenging of all genetic technologies. Here we reproduce the announcement of the first successful mammal clones, a Finnish Dorset lamb named Dolly as well as other less famous lambs. This advance was achieved by English embryologist Sir Ian Wilmut (1944–) and his team of biologists at Queen's Medical Research Institute at the University of Edinburgh. Their 1997 article in *Nature* describes the complex manipulation of cell nuclei, which had to be transferred from adult mammary cells into eggs that had their nuclei removed. The authors do not mention the prospects of human cloning, nor do they comment on the ethical implications of their revolutionary results.

* * *

Fertilization of mammalian eggs is followed by successive cell divisions and progressive differentiation, first into the early embryo and subsequently into all of the cell types that make up the adult animal. Transfer of a single nucleus at a specific stage of development, to an enucleated unfertilized egg, provided an opportunity to investigate whether cellular differentiation to that stage involved irreversible genetic modification. The first offspring to develop from a differentiated cell were born after nuclear transfer from an embryo-derived cell line that had been induced to become quiescent. Using the same procedure, we now report the birth of live lambs from three new cell populations established from adult mammary gland, fetus and embryo. The fact that a lamb was derived from an adult cell confirms that differentiation of that cell did not involve the irreversible modification of genetic material required for development to term. The birth of lambs from differentiated fetal and adult cells also reinforces previous speculation that by inducing donor cells to become quiescent it will be possible to obtain normal development from a wide variety of differentiated cells.

It has long been known that in amphibians, nuclei transferred from adult keratinocytcs established in culture support development to the juvenile, tadpole stage. Although this involves differentiation into complex tissues and organs, no development to the adult stage was reported, leaving open the question of whether a differentiated adult nucleus can be fully reprogrammed. Previously we reported the birth of live lambs after nuclear transfer from cultured embryonic cells that had been induced into quiescence. We suggested that inducing the donor cell to exit the growth phase causes changes in chromatin structure that facilitate reprogramming of gene expression and that development would be normal if nuclei are used from a variety of differentiated donor cells in similar regimes. Here we investigate whether normal development to term is possible when donor cells derived from fetal or adult tissue are induced to exit the growth cycle and enter the G0 phase of the cell cycle before nuclear transfer.

Three new populations of cells were derived from (1) a day-9 embryo, (2) a day-26 fetus and (3) mammary gland of a 6-year-old ewe in the last trimester of pregnancy. Morphology of the embryo-derived cells is unlike both mouse embryonic stem (ES) cells and the embryo-derived cells used in our previous study. Nuclear transfer was carried out according to one of our established protocols and reconstructed embryos transferred into recipient ewes. Ultrasound scanning detected 21 single fetuses on day 50–60 after oestrus. On subsequent scanning at ~ 14-day intervals, fewer fetuses were observed, suggesting either misdiagnosis or fetal loss. In total, 62% of fetuses were lost, a significantly greater proportion than the estimate of 6% after natural mating. Increased prenatal loss has been reported after embryo manipulation or culture of unreconstructed embryos. At about day 110 of pregnancy, four fetuses were dead, all from embryo-derived cells, and post-mortem analysis was possible after killing the ewes. Two fetuses had abnormal liver development, but no other abnormalities were detected and there was no evidence of infection.

Eight ewes gave birth to live lambs (Fig. 2). All three cell populations were represented. One weak lamb, derived from the fetal fibroblasts, weighed 3.1 kg and died within a few minutes of birth, although post-mortem analysis failed to find any abnormality or infection. At 12.5%, perinatal loss was not dissimilar to that occurring in a large study of commercial sheep, when 8% of lambs died within 24 h of birth. In all cases the lambs displayed the morphological characteristics of the breed used to derive the nucleus donors and not that of the oocyte donor. This alone indicates that the lambs could not have been born after inadvertent mating of either the oocyte donor or recipient ewes. In addition, DNA microsatellite analysis of the cell populations and the lambs at four polymorphic loci confirmed that each lamb was derived from the cell population used as nuclear donor. Duration of gestation is determined by fetal genotype, and in all cases gestation was longer than the breed mean. By contrast, birth weight is influenced by both maternal and fetal genotype. The birth weight of all lambs was within the range for single lambs born to Blackface ewes on our farm (up to 6.6 kg) and in most cases was within the range for the breed of the nuclear donor. There are no strict control observations for birth weight after embryo transfer between breeds, but the range in weight of lambs born to their own breed on our farms is 1.2–5.0 kg, 2–4.9 kg and 3–9 kg for the Finn Dorset, Welsh Mountain and Poll Dorset genotypes, respectively. The attainment of sexual maturity in the lambs is being monitored.

Development of embryos produced by nuclear transfer depends upon the maintenance of normal ploidy and creating the conditions for developmental regulation of gene expression. These responses are both influenced by the cell-cycle stage of donor and recipient cells and the interaction between them. A comparison of development of mouse and cattle embryos produced by nuclear transfer to oocytes or enucleated zygotes suggests that a greater proportion develop if the recipient is an oocyte. This may be because factors that bring about reprogramming of gene expression in a transferred nucleus are required for early development and are taken up by the pronuclei during development of the zygote.

If the recipient cytoplasm is prepared by enucleation of an oocyte at metaphase II, it is only possible to avoid chromosomal damage and maintain normal ploidy by transfer of diploid nuclei, but further experiments are required to define the optimum cell-cycle stage. Our studies with cultured cells suggest that there is an advantage if cells are quiescent. In earlier studies, donor cells were embryonic blastomeres that had not been induced into quiescence. Comparisons of the phases of the growth cycle showed that development was greater if donor cells were in mitosis or in the Gl phase of the cycle, rather than in S or G2 phases. Increased development using donor cells in G0, G1 or mitosis may reflect greater access for reprogramming factors present in the oocyte cycoplasm, but a direct comparison of these phases in the same cell population is required for a clearer understanding of the underlying mechanisms.

Together these results indicate that nuclei from a wide range of cell types should prove to be totipotent after enhancing opportunities for reprogramming by using appropriate combinations of these cell-cycle stages. In turn, the dissemination of the genetic improvement obtained within elite selection herds will be enhanced by limited replication of animals with proven performance by nuclear transfer from cells derived from adult animals. In addition, gene targeting in livestock should now be feasible by nuclear transfer from modified cell populations and will offer new opportunities in biotechnology. The techniques described also offer an opportunity to study the possible persistence and impact of epigenetic changes, such as imprinting and telomere shortening, which are known to occur in somatic cells during development and senescence, respectively.

The lamb born after nuclear transfer from a mammary gland cell is, to our knowledge, the first mammal to develop from a cell derived from an adult tissue. The phenotype of the donor cell is unknown. The primary culture contains mainly mammary epithelial (over 90%) as well as other differentiated cell types, including myoepithelial cells and fibroblasts. We cannot exclude the possibility that there is a small proportion of relatively undifferentiated stem cells able to support regeneration of the mammary gland during pregnancy. Birth of the lamb shows that during the development of that mammary cell there was no irreversible modification of genetic information required for development to term. This is consistent with the generally accepted view that mammalian differentiation is almost all achieved by systematic, sequential changes in gene expression brought about by interactions between the nucleus and the changing cytoplasmic environment.

Figure 2. Lamb number 6LL3 derived from the mammary gland of a Finn Dorset ewe with the Scottish Blackface ewe which was the recipient.

CHAPTER 25
EVOLUTION

*All life on Earth evolved from single celled organisms by the
process of natural selection.*

INTRODUCTION

*O*N THE *ORIGIN of Species* by Charles Darwin (1809–1882) is one of the most
original and influential books of all time. In 1859, when the first edition was
published, most scholars accepted the view that each species of animal and plant
was specially created. Darwin, by contrast, outlined a natural process by which all species
could arise over immense spans of time through the selection of individuals with favorable
traits. Like a skilled lawyer, Darwin leads his readers step by step through the argument.
Here we present excerpts from several chapters.

After a concise Introduction, which provides the historical context for his studies, Darwin
first reviews the dramatic changes that had been achieved through selective breeding of
animals and plants—notably domesticated pigeons (Chapter 1). He then outlines the three
main principles of his theory of evolution by natural selection: Each population of plants
and animals exhibits variations (Chapter 2); a "struggle for survival" occurs because many
more individuals are born than can survive (Chapter 3); and the process of natural selection
eliminates less fit individuals and leads to new species through gradual changes (Chapter 4).

Darwin concludes his volume (Chapter 14) with an oft-quoted philosophical statement,
"there is majesty in this view of life," that strives to blunt theological objections to his
theory.

On the Origin of Species by Means of Natural Selection, or the Preservation of Favoured Races in the Struggle for Life

by Charles Darwin (1859)

INTRODUCTION

WHEN ON BOARD H.M.S. Beagle, as naturalist, I was much struck with certain facts in the distribution of the inhabitants of South America, and in the geological relations of the present to the past inhabitants of that continent. These facts seemed to me to throw some light on the origin of species—that mystery of mysteries, as it has been called by one of our greatest philosophers. On my return home, it occurred to me, in 1837, that something might perhaps be made out on this question by patiently accumulating and reflecting on all sorts of facts which could possibly have any bearing on it. After five years' work I allowed myself to speculate on the subject, and drew up some short notes; these I enlarged in 1844 into a sketch of the conclusions, which then seemed to me probable: from that period to the present day I have steadily pursued the same object. I hope that I may be excused for entering on these personal details, as I give them to show that I have not been hasty in coming to a decision.

My work is now nearly finished; but as it will take me two or three more years to complete it, and as my health is far from strong, I have been urged to publish this Abstract. I have more especially been induced to do this, as Mr Wallace, who is now studying the natural history of the Malay archipelago, has arrived at almost exactly the same general conclusions that I have on the origin of species. Last year he sent to me a memoir on this subject, with a request that I would forward it to Sir Charles Lyell, who sent it to the Linnean Society, and it is published in the third volume of the journal of that Society. Sir C. Lyell and Dr Hooker, who both knew of my work—the latter having read my sketch of 1844—honoured me by thinking it advisable to publish, with Mr Wallace's excellent memoir, some brief extracts from my manuscripts.

This Abstract, which I now publish, must necessarily be imperfect. I cannot here give references and authorities for my several statements; and I must trust to the reader reposing some confidence in my accuracy. No doubt errors will have crept in, though I hope I have always been cautious in trusting to good authorities alone. I can here give only the general conclusions at which I have arrived, with a few facts in illustration, but which, I hope, in most cases will suffice. No one can feel more sensible than I do of the necessity of hereafter publishing in detail all the facts, with references, on which my conclusions have been grounded; and I hope in a future work to do this. For I am well aware that scarcely a single point is discussed in this volume on which facts cannot be adduced, often apparently leading to conclusions directly opposite to those at which I have arrived. A fair result can be obtained only by fully stating and balancing the facts and arguments on both sides of each question; and this cannot possibly be here done.

CHAPTER 1—VARIATION UNDER DOMESTICATION
... On the Breeds of the Domestic Pigeon.

Believing that it is always best to study some special group, I have, after deliberation, taken up domestic pigeons. I have kept every breed which I could purchase or obtain, and have been most kindly favoured with skins from several quarters of the world, more especially by the Hon. W. Elliot from India, and by the Hon. C. Murray from Persia. Many treatises in different languages have been published on pigeons, and some of them are very important, as being of considerably antiquity. I have associated with several eminent fanciers, and have been permitted to join two of the London Pigeon Clubs. The diversity of the breeds is something astonishing. Compare the English carrier and the short-faced tumbler, and see the wonderful difference in their beaks, entailing corresponding differences in their skulls. The carrier, more especially the male bird, is also remarkable from the wonderful development of the carunculated skin about the head, and this is accompanied by greatly elongated eyelids, very large external orifices to the nostrils, and a wide gape of mouth. The short-faced tumbler has a beak in outline almost like that of a finch; and the common tumbler has the singular and strictly inherited habit of flying at a great height in a compact flock, and tumbling in the air head over heels. The runt is a bird of great size, with long, massive beak and large feet; some of the sub-breeds of runts have very long necks, others very long wings and tails, others singularly short tails. The barb is allied to the carrier, but, instead of a very long beak, has a very short and very broad one. The pouter has a much elongated body, wings, and legs; and its enormously developed crop, which it glories in inflating, may well excite astonishment and even laughter. The turbit has a very short and conical beak, with a line of reversed feathers down the breast; and it has the habit of continually expanding slightly the upper part of the oesophagus. The Jacobin has the feathers so much reversed along the back of the neck that they form a hood, and it has, proportionally to its size, much elongated wing and tail feathers. The trumpeter and laugher, as their names express, utter a very different coo from the other breeds. The fantail has thirty or even forty tail-feathers, instead of twelve or fourteen, the normal number in all members of the great pigeon family; and these feathers are kept expanded, and are carried so erect that in good birds the head and tail touch; the oil-gland is quite aborted. Several other less distinct breeds might have been specified.

In the skeletons of the several breeds, the development of the bones of the face in length and breadth and curvature differs enormously. The shape, as well as the breadth and length of the ramus of the lower jaw, varies in a highly remarkable manner. The number of the caudal and sacral vertebrae vary; as does the number of the ribs, together with their relative breadth and the presence of processes. The size and shape of the apertures in the sternum are highly variable; so is the degree of divergence and relative size of the two arms of the furcula. The proportional width of the gape of mouth, the proportional length of the eyelids, of the orifice of the nostrils, of the tongue (not always in strict correlation with the length of beak), the size of the crop and of the upper part of the oesophagus; the development and abortion of the oil-gland; the number of the primary wing and caudal feathers; the relative length of wing and tail to each other and to the body; the relative length of leg and of the feet; the number of scutellae on the toes, the development of skin between the toes, are all points of structure which are variable. The period at which the perfect plumage is acquired varies, as does the state of the down with which the nestling birds are clothed when hatched. The shape and size of the eggs vary. The manner of flight differs remarkably; as does in some breeds the voice and disposition. Lastly, in certain breeds, the males and females have come to differ to a slight degree from each other.

Altogether at least a score of pigeons might be chosen, which if shown to an ornithologist, and he were told that they were wild birds, would certainly, I think, be ranked by him as well-defined species. Moreover, I do not believe that any ornithologist would place the English carrier, the short-faced tumbler, the runt, the barb, pouter, and fantail in the same genus; more especially as in each of these breeds several truly-inherited sub-breeds, or species as he might have called them, could be shown him.

Great as the differences are between the breeds of pigeons, I am fully convinced that the common opinion of naturalists is correct, namely, that all have descended from the rock-pigeon (Columba livia), including under this term several geographical races or sub-species, which differ from each other in the most trifling respects. As several of the reasons which have led me to this belief are in some degree applicable in other cases, I will here briefly give them. If the several breeds are not varieties, and have not proceeded from the rock-pigeon, they must have descended from at least seven or eight aboriginal stocks; for it is impossible to make the present domestic breeds by the crossing of any lesser number: how, for instance, could a pouter be produced by crossing two breeds unless one of the parent-stocks possessed the characteristic enormous crop? The supposed aboriginal stocks must all have been rock-pigeons, that is, not breeding or willingly perching on trees. But besides C. livia, with its geographical sub-species, only two or three other species of rock-pigeons are known; and these have not any of the characters of the domestic breeds. Hence the supposed aboriginal stocks must either still exist in the countries where they were originally domesticated, and yet be unknown to ornithologists; and this, considering their size, habits, and remarkable characters, seems very improbable; or they must have become extinct in the wild state. But birds breeding on precipices, and good fliers, are unlikely to be exterminated; and the common rock-pigeon, which has the same habits with the domestic breeds, has not been exterminated even on several of the smaller British islets, or on the shores of the Mediterranean. Hence the supposed extermination of so many species having similar habits with the rock-pigeon seems to me a very rash assumption. Moreover, the several above-named domesticated breeds have been transported to all parts of the world, and, therefore, some of them must have been carried back again into their native country; but not one has ever become wild or feral, though the dovecot-pigeon, which is the rock-pigeon in a very slightly altered state, has become feral in several places. Again, all recent experience shows that it is most difficult to get any wild animal to breed freely under domestication; yet on the hypothesis of the multiple origin of our pigeons, it must be assumed that at least seven or eight species were so thoroughly domesticated in ancient times by half-civilized man, as to be quite prolific under confinement.

An argument, as it seems to me, of great weight, and applicable in several other cases, is, that the above-specified breeds, though agreeing generally in constitution, habits, voice, colouring, and in most parts of their structure, with the wild rock-pigeon, yet are certainly highly abnormal in other parts of their structure: we may look in vain throughout the whole great family of Columbidae for a beak like that of the English carrier, or that of the short-faced tumbler, or barb; for reversed feathers like those of the jacobin; for a crop like that of the pouter; for tail-feathers like those of the fantail. Hence it must be assumed not only that half-civilized man succeeded in thoroughly domesticating several species, but that he intentionally or by chance picked out extraordinarily abnormal species; and further, that these very species have since all become extinct or unknown. So many strange contingencies seem to me improbable in the highest degree.

Some facts in regard to the colouring of pigeons well deserve consideration. The rock-pigeon is of a slaty-blue, and has a white rump (the Indian sub-species, C. intermedia of Strickland, having it bluish); the tail has a terminal dark bar, with the bases of the outer feathers externally edged with white; the wings have two black bars: some semi-domestic breeds and some apparently truly wild breeds have, besides the two black bars, the wings chequered with black. These several marks do not occur together in any other species of the whole family. Now, in every one of the domestic breeds, taking thoroughly well-bred birds, all the above marks, even to the white edging of the outer tail-feathers, sometimes concur perfectly developed. Moreover, when two birds belonging to two distinct breeds are crossed, neither of which is blue or has any of the above-specified marks, the mongrel offspring are very apt suddenly to acquire these characters; for instance, I crossed some uniformly white fantails with some uniformly black barbs, and they produced mottled brown and black birds; these I again crossed together, and one grandchild of the pure white fantail and pure black barb was of as beautiful a blue colour, with the white rump, double black wing-bar, and barred and white-edged tail-feathers, as any wild rock-pigeon! We can understand these facts, on the well-known principle of reversion to ancestral characters, if all the domestic breeds have descended from the rock-pigeon. But if we deny this, we must make one of the two following highly improbable suppositions. Either, firstly, that all the several imagined aboriginal stocks were coloured and marked like the rock-pigeon, although no other existing species is thus coloured and marked, so that in each separate breed there might be a tendency to revert to the very same colours and markings. Or, secondly, that each breed, even the purest, has within a dozen or, at most, within a score of generations, been crossed by the rock-pigeon: I say within a dozen or twenty generations, for we know of no fact countenancing the belief that the child ever reverts to some one ancestor, removed by a greater number of generations. In a breed which has been crossed only once with some distinct breed, the tendency to reversion to any character derived from such cross will naturally become less and less, as in each succeeding generation there will be less of the foreign blood; but when there has been no cross with a distinct breed, and there is a tendency in both parents to revert to a character, which has been lost during some former generation, this tendency, for all that we can see to the contrary, may be transmitted undiminished for an indefinite number of generations. These two distinct cases are often confounded in treatises on inheritance.

Lastly, the hybrids or mongrels from between all the domestic breeds of pigeons are perfectly fertile. I can state this from my own observations, purposely made on the most distinct breeds. Now, it is difficult, perhaps impossible, to bring forward one case of the hybrid offspring of two animals *clearly distinct* being themselves perfectly fertile. Some authors believe that long-continued domestication eliminates this strong tendency to sterility: from the history of the dog I think there is some probability in this hypothesis, if applied to species closely related together, though it is unsupported by a single experiment. But to extend the hypothesis so far as to suppose that species, aboriginally as distinct as carriers, tumblers, pouters, and fantails now are, should yield offspring perfectly fertile, *inter se*, seems to me rash in the extreme.

From these several reasons, namely, the improbability of man having formerly got seven or eight supposed species of pigeons to breed freely under domestication; these supposed species being quite unknown in a wild state, and their becoming nowhere feral; these species having very abnormal characters in certain respects, as compared with all other Columbidae, though so like in most other respects to the rock-pigeon; the blue colour and various marks occasionally appearing in all the breeds, both when kept pure and when crossed; the mongrel offspring being perfectly fertile; from these several reasons, taken together, I can feel no doubt that all our domestic breeds have descended from the Columba livia with its geographical sub-species.

In favour of this view, I may add, firstly, that C. livia, or the rock-pigeon, has been found capable of domestication in Europe and in India; and that it agrees in habits and in a great number of points of structure with all the domestic breeds. Secondly, although an English carrier or short-faced tumbler differs immensely in certain characters from the rock-pigeon, yet by comparing the several sub-breeds of these breeds, more especially those brought from distant countries, we can make an almost perfect series between the extremes of structure.

Thirdly, those characters which are mainly distinctive of each breed, for instance the wattle and length of beak of the carrier, the shortness of that of the tumbler, and the number of tail-feathers in the fantail, are in each breed eminently variable; and the explanation of this fact will be obvious when we come to treat of selection. Fourthly, pigeons have been watched, and tended with the utmost care, and loved by many people. They have been domesticated for thousands of years in several quarters of the world; the earliest known record of pigeons is in the fifth Aegyptian dynasty, about 3000 B.C., as was pointed out to me by Professor Lepsius; but Mr. Birch informs me that pigeons are given in a bill of fare in the previous dynasty. In the time of the Romans, as we hear from Pliny, immense prices were given for pigeons; "nay, they are come to this pass, that they can reckon up their pedigree and race." Pigeons were much valued by Akber Khan in India, about the year 1600; never less than 20,000 pigeons were taken with the court. 'The monarchs of Iran and Turan sent him some very rare birds;' and, continues the courtly historian, 'His Majesty by crossing the breeds, which method was never practised before, has improved them astonishingly.' About this same period the Dutch were as eager about pigeons as were the old Romans. The paramount importance of these considerations in explaining the immense amount of variation which pigeons have undergone, will be obvious when we treat of Selection. We shall then, also, see how it is that the breeds so often have a somewhat monstrous character. It is also a most favourable circumstance for the production of distinct breeds, that male and female pigeons can be easily mated for life; and thus different breeds can be kept together in the same aviary.

CHAPTER 2—VARIATION UNDER NATURE

Before applying the principles arrived at in the last chapter to organic beings in a state of nature, we must briefly discuss whether these latter are subject to any variation. To treat this subject at all properly, a long catalogue of dry facts should be given; but these I shall reserve for my future work. Nor shall I here discuss the various definitions which have been given of the term species. No one definition has as yet satisfied all naturalists; yet every naturalist knows vaguely what he means when he speaks of a species. Generally the term includes the unknown element of a distinct act of creation. The term 'variety' is almost equally difficult to define; but here community of descent is almost universally implied, though it can rarely be proved. We have also what are called monstrosities; but they graduate into varieties. By a monstrosity I presume is meant some considerable deviation of structure in one part, either injurious to or not useful to the species, and not generally propagated. Some authors use the term 'variation' in a technical sense, as implying a modification directly due to the physical conditions of life; and 'variations' in this sense are supposed not to be inherited: but who can say that the dwarfed condition of shells in the brackish waters of the Baltic, or dwarfed plants on Alpine summits, or the thicker fur of an animal from far northwards, would not in some cases be inherited for at least some few generations? and in this case I presume that the form would be called a variety.

Again, we have many slight differences which may be called individual differences, such as are known frequently to appear in the offspring from the same parents, or which may be presumed to have thus arisen, from being frequently observed in the individuals of the same species inhabiting the same confined locality. No one supposes that all the individuals of the same species are cast in the very same mould. These individual differences are highly important for us, as they afford materials for natural selection to accumulate, in the same manner as man can accumulate in any given direction individual differences in his domesticated productions. These individual differences generally affect what naturalists consider unimportant parts; but I could show by a long catalogue of facts, that parts which must be called important, whether viewed under a physiological or classificatory point of view, sometimes vary in the individuals of the same species.

CHAPTER 3—STRUGGLE FOR EXISTENCE

Before entering on the subject of this chapter, I must make a few preliminary remarks, to show how the struggle for existence bears on Natural Selection. ... Again, it may be asked, how is it that varieties, which I have called incipient species, become ultimately converted into good and distinct species, which in most cases obviously differ from each other far more than do the varieties of the same species? How do those groups of species, which constitute what are called distinct genera, and which differ from each other more than do the species of the same genus, arise? All these results, as we shall more fully see in the next chapter, follow inevitably from the struggle for life. Owing to this struggle for life, any variation, however slight and from whatever cause proceeding, if it be in any degree profitable to an individual of any species, in its infinitely complex relations to other organic beings and to external nature, will tend to the preservation of that individual, and will generally be inherited by its offspring. The offspring, also, will thus have a better chance of surviving, for, of the many individuals of any species which are periodically born, but a small number can survive. I have called this principle, by which each slight variation, if useful, is preserved, by the term of Natural Selection, in order to mark its relation to man's power of selection. We have seen that man by selection can certainly produce great results, and can adapt organic beings to his own uses, through the accumulation of slight but useful variations, given to him by the hand of Nature. But Natural Selection, as we shall hereafter see, is a power incessantly ready for action, and is as immeasurably superior to man's feeble efforts, as the works of Nature are to those of Art. ...

I should premise that I use the term Struggle for Existence in a large and metaphorical sense, including dependence of one being on another, and including (which is more important) not only the life of the individual, but success in leaving progeny. ... A struggle for existence inevitably follows from the high rate at which all organic beings tend to increase. Every being, which during its natural lifetime produces several eggs or seeds, must suffer destruction during some period of its life, and during some season or occasional year, otherwise, on the principle of geometrical increase, its numbers would quickly become so inordinately great that no country could support the product. Hence, as more individuals are produced than can possibly survive, there must in every case be a struggle for existence, either one individual with another of the same species, or with the individuals of distinct species, or with the physical conditions of life. It is the doctrine of Malthus applied with manifold force to the whole animal and vegetable kingdoms; for in this case there can be no artificial increase of food, and no prudential restraint from marriage. Although some species may be now increasing, more or less rapidly, in numbers, all cannot do so, for the world would not hold them.

There is no exception to the rule that every organic being naturally increases at so high a rate, that if not destroyed, the earth would soon be covered by the progeny of a single pair. Even slow-breeding man has doubled in twenty-five years, and at this rate, in a few thousand years, there would literally not be standing room for his progeny. Linnaeus has calculated that if an annual plant produced only two seeds and there is no plant so unproductive as this and their seedlings next year produced two, and so on, then in twenty years there would be a million plants. The elephant is reckoned to be the slowest breeder of all known animals, and I have taken some pains to estimate its probable minimum rate of natural increase: it will be under the mark to assume that it breeds when thirty years old, and goes on breeding till ninety years old, bringing forth three pairs of young in this interval; if this be so, at the end of the fifth century there would be alive fifteen million elephants, descended from the first pair.

CHAPTER 4—NATURAL SELECTION

How will the struggle for existence, discussed too briefly in the last chapter, act in regard to variation? Can the principle of selection, which we have seen is so potent in the hands of man, apply in nature? I think we shall see that it can act most effectually. Let it be borne in mind in what an endless number of strange peculiarities our domestic productions, and, in a lesser degree, those under nature, vary; and how strong the hereditary tendency is. Under domestication, it may be truly said that the, whole organisation becomes in some degree plastic. Let it be borne in mind how infinitely complex and close-fitting are the mutual relations of all organic beings to each other and to their physical conditions of life. Can it, then, be thought improbable, seeing that variations useful to man have undoubtedly occurred, that other variations useful in some way to each being in the great and complex battle of life, should sometimes occur in the course of thousands of generations? If such do occur, can we doubt (remembering that many more individuals are born than can possibly survive) that individuals having any advantage, however slight, over others, would have the best chance of surviving and of procreating their kind? On the other hand, we may feel sure that any variation in the least degree injurious would be rigidly destroyed. This preservation of favourable variations and the rejection of injurious variations, I call Natural Selection. Variations neither useful nor injurious would not be affected by natural selection, and would be left a fluctuating element, as perhaps we see in the species called polymorphic. ...

As man can produce and certainly has produced a great result by his methodical and unconscious means of selection, what may not nature effect? Man can act only on external and visible characters: nature cares nothing for appearances, except in so far as they may be useful to any being. She can act on every internal organ, on every shade of constitutional difference, on the whole machinery of life. Man selects only for his own good; Nature only for that of the being which she tends. Every selected character is fully exercised by her; and the being is placed under well-suited conditions of life. Man keeps the natives of many climates in the same country; he seldom exercises each selected character in some peculiar and fitting manner; he feeds a long and a short beaked pigeon on the same food; he does not exercise a long-backed or long-legged quadruped in any peculiar manner; he exposes sheep with long and short wool to the same climate. He does not allow the most vigorous males to struggle for the females. He does not rigidly destroy all inferior animals, but protects during each varying season, as far as lies in his power, all his productions. He often begins his selection by some half-monstrous form; or at least by some modification prominent enough to catch his eye, or to be plainly useful to him. Under nature, the slightest difference of structure or constitution may well turn the nicely-balanced scale in the struggle for life, and so be preserved. How fleeting are the wishes and efforts of man! how short his time! and consequently how poor will his products be, compared with those accumulated by nature during whole geological periods. Can we wonder, then, that nature's productions should be far 'truer' in character than man's productions; that they should be infinitely better adapted to the most complex conditions of life, and should plainly bear the stamp of far higher workmanship?

It may be said that natural selection is daily and hourly scrutinising, throughout the world, every variation, even the slightest; rejecting that which is bad, preserving and adding up all that is good; silently and insensibly working, whenever and wherever opportunity offers, at the improvement of each organic being in relation to its organic and inorganic conditions of life. We see nothing of these slow changes in progress, until the hand of time has marked the long lapses of ages, and then so imperfect is our view into long past geological ages, that we only see that the forms of life are now different from what they formerly were.

CHAPTER 14—RECAPITULATION AND CONCLUSION ...

Authors of the highest eminence seem to be fully satisfied with the view that each species has been independently created. To my mind it accords better with what we know of the laws impressed on matter by the Creator, that the production and extinction of the past and present inhabitants of the world should have been due to secondary causes, like those determining the birth and death of the individual. When I view all beings not as special creations, but as the lineal descendants of some few beings which lived long before the first bed of the Silurian system was deposited, they seem to me to become ennobled. Judging from the past, we may safely infer that not one living species will transmit its unaltered likeness to a distant futurity. And of the species now living very few will transmit progeny of any kind to a far distant futurity; for the manner in which all organic beings are grouped, shows that the greater number of species of each genus, and all the species of many genera, have left no descendants, but have become utterly extinct. We can so far take a prophetic glance into futurity as to foretell that it will be the common and widely-spread species, belonging to the larger and dominant groups, which will ultimately prevail and procreate new and dominant species. As all the living forms of life are the lineal descendants of those which lived long before the Silurian epoch, we may feel certain that the ordinary succession by generation has never once been broken, and that no cataclysm has desolated the whole world. Hence we may look with some confidence to a secure future of equally inappreciable length. And as natural selection works solely by and for the good of each being, all corporeal and mental endowments will tend to progress towards perfection.

It is interesting to contemplate an entangled bank, clothed with many plants of many kinds, with birds singing on the bushes, with various insects flitting about, and with worms crawling through the damp earth, and to reflect that these elaborately constructed forms, so different from each other, and dependent on each other in so complex a manner, have all been produced by laws acting around us. These laws, taken in the largest sense, being Growth with Reproduction; inheritance which is almost implied by reproduction; Variability from the indirect and direct action of the external conditions of life, and from use and disuse; a Ratio of Increase so high as to lead to a Struggle for Life, and as a consequence to Natural Selection, entailing Divergence of Character and the Extinction of less-improved forms. Thus, from the war of nature, from famine and death, the most exalted object which we are capable of conceiving, namely, the production of the higher animals, directly follows. There is grandeur in this view of life, with its several powers, having been originally breathed into a few forms or into one; and that, whilst this planet has gone cycling on according to the fixed law of gravity, from so simple a beginning endless forms most beautiful and most wonderful have been, and are being, evolved.

ACKNOWLEDGMENTS

Alessandro Volta, from "On the Electricity Excited by the Mere Contact of Conducting Substances of Different Kinds" in *A Source Book in Physics*, ed. William Francis Magie, pp. 427–431. Copyright © 1935 by McGraw-Hill Book Company, Inc. Reprinted with permission.

Isaac Newton, from "Mathematical Principles of Natural Philosophy" in *A Source Book in Physics*, ed. William Francis Magie, pp. 31–38. Copyright © 1935 by McGraw-Hill Book Company, Inc. Reprinted with permission.

Michael Faraday, from "Experimental Researches in Electricity" in *A Source Book in Physics*, ed. William Francis Magie, pp. 478–485. Copyright © 1935 by McGraw-Hill Book Company, Inc. Reprinted with permission.

Heinrich Hertz, from "Electric radiation" in *A Source Book in Physics*, ed. William Francis Magie, pp. 550–559. Copyright © 1935 by McGraw-Hill Book Company, Inc. Reprinted with permission.

Graham P. Collins, from "Large Hadron Collider: The Discovery Machine" in *Scientific American*, January 17, 2008. Copyright © 2008 by Scientific American Inc. Reprinted with permission.

Bruce C. Heezen, Marie Tharp, and Maurice Ewing, from "The Floors of the Ocean," in Geological Society of America, p. 122. Copyright © 1959 by The Geological Society of America, Inc. Reprinted with permission.

Frederick J. Vine and Drummond H. Matthews, from "Magnetic Anomalies Over Ocean Ridges" in *Nature*, pp. 947–949. Copyright © 1963 by *Nature*. Reprinted with permission.

James Watson and Francis Crick, from "A Structure for Deoxyribose Nucleic Acid" in *Nature*, pp. 737–738. Copyright © 1953 by *Nature*. Reprinted with permission.

Ian Wilmut and colleagues, from "Viable Offspring Derived from Fetal and Adult Mammalian Cells" in *Nature*, pp. 810–813. Copyright © 1997 by *Nature*. Reprinted with permission.

Robert H. MacArthur and Edward O. Wilson, from "An Equilibrium Theory of Insular Zoogeography" in *Evolution*, pp. 373–387. Copyright © 1963 by The Society for the Study of Evolution. Reprinted with permission.

Randall K. Saiki and colleagues, "Enzymatic Amplification of β-globin Genomic Sequences and Restriction Site Analysis for Diagnosis of Sickle Cell Anemia" in *Science 230*, pp. 1350–1354. Copyright © 1985 by the American Association for the Advancement of Science. Reprinted with permission.

All other selections in this anthology copyright in the public domain.

LaVergne, TN USA
23 December 2009
167814LV00005B/1/P

9 781935 551126